Lecture Notes
in Control and Information Sciences 234

Editor: M. Thoma

Springer-Verlag London Ltd.

Paolo Arena, Luigi Fortuna, Giovanni Muscato
and Maria G. Xibilia

Neural Networks in Multidimensional Domains

Fundamentals and New Trends in Modelling and Control

 Springer

Authors

Dr Paolo Arena · Dr Luigi Fortuna
Dr Giovanni Muscato · Dr Maria Gabriella Xibilia
Università di Catania, D E E S, Via Andrea Doria 6, 95125 Catania, Italia

ISBN 978-1-85233-006-4

British Library Cataloguing in Publication Data
Neural networks in multidimensional domains : fundamentals
   and new trends in modelling and control. -   (Lecture notes
   in control and information sciences ; 234)
   1.Neural networks (Computer science)
   I.Arena, Paolo
   006.3'2
ISBN 978-1-85233-006-4
Library of Congress Cataloging-in-Publication Data
Neural networks in multidimensional domains : fundamentals and new
   trends in modelling and control / Paolo Arena ... [et al.].
      p.   cm. - - (Lecture notes in control and information sciences ; 234)
   Includes index.
   ISBN 978-1-85233-006-4   ISBN 978-1-84628-527-1 (eBook)
   DOI 10.1007/978-1-84628-527-1
   1. Neural networks (Computer science) 2. Computer simulation.
   3. Automatic control. I. Arena, Paolo, 1966-   . II. Series.
   QA76.87.N4873   1998   98-6387
   006.3'2- -dc21   CIP

Typesetting: Camera ready by authors

69/3830-543210 Printed on acid-free paper

Preface

A growing interest has been shown in the last few years in the field of neural networks, and a number of books have been published. The topics dealt with in these books cover a wide range. Various types of neural networks have been widely discussed and much effort has been devoted both to developing the theoretical aspects and to producing significant applications. Recently, efforts have focused on presenting the various results in a unified framework. In this monograph a new subject regarding neural networks in multi-dimensional domains is approached. The central question addressed is a new generalization of multi-layer perceptrons in complex, vectorial and hypercomplex domains. The approximation capabilities of these networks and the learning algorithms will be discussed in a multi-dimensional context.

The book includes the theoretical basis for characterizing neural networks in the hypercomplex domain, applications referring to attractive themes in system engineering and a Matlab software tool for using quaternion algebra and hypercomplex neural networks.

The purpose of this book is to give the reader a view, in an immediate form, of the new area of hypercomplex neural networks, to show their applicability in some advanced engineering topics, to encourage him to make experiments, using the Matlab tool, after having understood the main guidelines of the subject.

The appropriate background for this text is the fundamentals of neural networks. The manuscript is intended as a research report, but a great effort has been made to make the subject comprehensible for graduate students in computer engineering, control engineering, computer sciences, mathematics, physics and related disciplines.

The monograph is organized as follows:

an introduction to the classical multi-layer perceptron (MLP) neural network is given in Chapter 1 together with some notes on the neural approach to the problem of identifying non linear systems. The generalization of the MLP in the complex domain is the central topic of Chapter 2, where the Complex Multi-layer Perceptron (CMLP) is described together with some theoretical results on its approximation capabilities, and learning algorithm is presented. A first class of multi-dimensional neural networks is introduced in Chapter 3 where vectorial neural networks are dealt with. After having introduced the basic tools of Quaternion algebra in Chapter 4, the MLP in quaternion algebra is considered in detail in Chapter 5, where the key points of the Hypercomplex Multylayer Perceptron are presented. In Chapter 6 and Chapter 7 selected applications of the presented theory are described. More specifically, in Chapter 6 the problem of Chaotic time series prediction using the CMLP and HMLP neural networks is discussed and the improvement with respect to classical prediction strategies is shown. A neural controller based on the HMLP, is presented in Chapter 7 with particular reference to robotics. In Appendix A some details regarding Quaternion Algebra are given. Appendix B includes the principles of chaotic circuits. The Matlab software tool developed for quaternion operations and hypercomplex neural network implementation is described in Appendix C. The files can be directly requested to the authors by E-mail at the address parena@dees.unict.it

Contents

List of Figures

Notations

Notations and Symbols

$\Re$:Real algebra;

$\Re^n$:space spanned by n real component vectors;

$\Re^{n*m}$:space spanned by real matrices with n rows and m columns;

C :complex algebra;

C^n :space spanned by n complex component vectors;

H :quaternion algebra;

H^n :space spanned by n quaternion component vectors;

$C^0(X)$:space of continuous functions defined on the set X;

I_n :n dimension unit cube $[0\ 1]^n$;

D_n :n dimension disc in C;

$*$:operator which cojugates a complex or a quaternion number;

T :operator which transposes a matrix or a vector;

$\times$:vector product;

$\cdot$:scalar product;

$\odot$:product component by component between complex or quaternion numbers;

$\otimes$:product between two quaternions;

o :function composition;

$|\bullet|$: modulus;

$\|\bullet\|$: norm (see definition in Appendix);

MLP: Multi-Layer Perceptron;

CMLP: Complex Multi-Layer Perceptron;

HMLP: Hypercomplex Multi-Layer Perceptron;

BP : Back-Propagation.

Chapter 1

Introduction to MLP neural networks

The brain has always been the most perfect and well-organized processing structure, and its emulation still represents one of the most appealing challenges for present and future research. With this statement we do not want to detract from the real revolution that modern computers have performed, but it is well known that there exist some tasks, such as pattern recognition and restoration, object classification or abstract scheme learning, that can hardly be implemented on conventional computers, while our brain is able to perform them in real time. Taking into consideration the fact that typical biological processing times are several orders of magnitude greater than those of the simplest conventional computing machine, it is apparent that such performance is due to the structure of the brain. It is this idea that has stimulated scientists in the last 50 years to pay attention to the processing philosophy of the brain and to try to emulate its characteristics by simulating its structure. The brain processes information in a highly parallel and distributed fashion; in a sense its behaviour can be viewed as a large set of asynchronous, chemically drived, pulse generating switches. Massive cooperation between a large number of simple processing elements (neurons) gives rise to a number of very interesting properties, which do not have a counterpart in traditional computing structures.

These properties include:

- *robustness and fault tolerance* : several neurons die every day, without greatly affecting the efficiency of the brain;

- *flexibility* : it possesses a great capacity for adaptation which derives from its ability to *learn from the environment*;

- *data fusion capability*: it is able to deal concurrently with different types of information, since it has the ability to *extract rules* from environmental clues;

- it does not require information described with high numerical precision; it can also be fuzzy or noise-corrupted.

These are the main reasons that stimulated the growth of *connectionism*, i.e. the part of artificial intelligence devoted to building models of *thinking machines* drawing information from neurobiological sciences. The idea is to transfer the available knowledge about the functional characteristics of the human brain into algorithmic models, and then to build neurocomputers.

Even if the inspiration for artificial neural networks is undoubtedly drawn from biology, neural network theory has been mixed with traditional strategies, methodologies and algorithms coming from systems engineering. In a sense, it can be said that certain techniques, sometimes very old ones, have been fitted into new architectures, thus gaining new life.

From 1958, when *Rosenblatt* introduced the *Perceptron*, to today, there has been a never-ending development of theoretical studies and applications of artificial neural systems. One of the most important events in the field was the introduction of the *Multi-Layer Perceptron* structure (MLP) in 1986, which was able to perform hard nonlinear mapping and classification, by using a powerful learning algorithm, *Back Propagation*. From then on new applications in all scientific fields never ceased to appear, together with novel theoretical results on existing neural architectures, as well as the introduction of innovative networks. Neural processing has gained a fundamental role in the following fields:

- *classification* : neural networks are able to perform clustering operations on different classes starting from the individuals belonging to the classes;

- *noise filtering* : an artificial neural network can be trained to recognise signals or images. Suitable learning can improve its recognition capabilities even in the presence of a high level of noise corruption;

- *information processing* : neural systems are powerful devices which can pick out information from available measures, even in difficult cases of noise corruption, with performance levels that are often very close to those an expert can achieve;

- *prediction* : this problem is often encountered when it is necessary to estimate the future trend of certain variables based on their past behaviour. It is strictly related to the problem of system modelling and identification. In this field neural networks play a fundamental role, in that they use some of the algorithms typical of traditional estimation techniques, but with suitable modifications.

The capacity of neural networks to solve these problems is based on fundamental theoretical results, introduced in literature a few years after the introduction of the MLP. The input-output mapping produced by the MLP can be seen as a linear combination of particular analytic, real valued functions that are *dense* in the set of continuous, real valued functions defined in a compact set of R^n. Therefore every function belonging to such a set can be uniformly approximated with the desired index of accuracy. If it is also considered that the same input-output relation, by properly selecting the network inputs and output, constitutes a model structure useful for non-linear system modelling, the neural network approach acquires more relevance. All of these problems will be addressed in some detail in this chapter, and their extensions to the multidimensional numerical domains, the complex and the hypercomplex set, will be presented in the rest of the book. The aim of the discussion so far is to briefly introduce some of the main characteristics of neural structures that make them useful devices in several practical cases. Of course, even if a certain number of powerful theoretical results have been obtained, several open problems still remain without an answer, including, for instance, a direct methodology for architectural design as a function of the problem to be dealt with. A characteristic of neural computation is that it involves a great number of variables: redundancy is therefore a common denominator, and only heuristic criteria can help in the right topology design, based on certain indexes of the behaviour of the network. However, if from the engineering point of view it is fundamental to try to minimize the number of parameters involved in the neural network, on the other hand the source of inspiration must not be forgotten: the power of the brain lies in the parallel and distributed way information is processed by a huge number of mutually interacting neurons: redundancy is a key characteristic which implies adaptation, fault tolerance and robustness. Therefore, neural network research is the result of a compromise between two different ways of thinking: the biological and the engineering one. It is from the synergy between these two scientific fields that neural network research has received its best results. In fact, if some open questions still remain in connectionism, several functional aspects and models of massive cooperation among living neurons and a great number of anatomico-functional details still remain unsolved. From this point of view, close collaboration implies mutual advantages.

One of the aims of this book is to introduce a natural way to reduce the number of parameters involved in a neural structure by exploiting the properties of multidimensional numerical domains. In the following chapters some results will be introduced concerning suitable neural structures, defined in complex and hypercomplex algebras, which are able to achieve the same approximation capabilities as real valued MLPs, i.e. traditional neural systems, but which employ a smaller number of real parameters. All of the structures that will be discussed are a generalization of the well-known MLP. Therefore, in the following chapter, a brief introduction to classical MLPs will be presented.

1.1   The Multi-Layer Perceptron

The Multi-Layer Perceptron (MLP) is undoubtedly the most widely used neural network structure, in relation to the great number of applications developed in various scientific fields. It was introduced by Rumelhart and McClelland [13] to overcome the drawbacks shown by the *single layer perceptron* introduced by Rosenblatt. The latter was, in fact, only able to solve linearly separable problems, i.e. tasks involving decision regions separable by hyperplanes, and not more complex curves. The introduction of MLPs increased the importance of neural processing. The presence of hidden units, arranged in one or more layers, performs a kind of recoding action on the information flowing through the input layer, and therefore allows even very difficult associations to be made between input and output patterns. Moreover, the approximation capabilities of the activation functions that can be embedded in the hidden neurons allow arbitrary decision regions to be obtained. The recoding action of the hidden units is made possible by a powerful learning algorithm, that allows information coming from the outside to be propagated and coded inside the network up to the hidden units.

The well-known MLP structure is shown in Fig.1.1.

It is made up of a certain number of *units* or *neurons*, arranged in layers, and

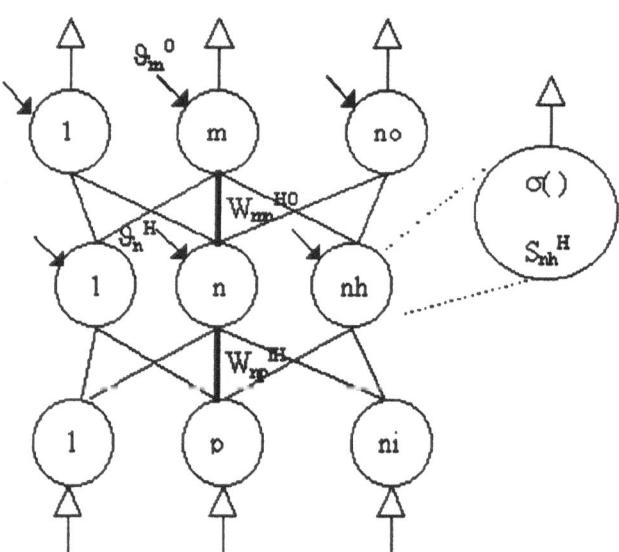

Figure 1.1: *MLP with one hidden layer*

connected with each other through links whose values determine the *strength* or *weight* of the connections themselves, and therefore the degree of influence one neuron has over another one. For these reasons the connections among the MLP neurons are also called *weights*. In the MLP structure, each neuron in a layer is connected to all of the units belonging to the following layer and to all of the units of the preceding layer. However, there are no weights among units in the same layer, nor bridging layers. Once the network topology has been fixed, the connection structure represents the site of the network's knowledge; in fact, at the end of the learning process, the weight value represents what the network has learnt. From this point of view the MLP closely resembles the living neural structure: memory is distributed in the connections among neurons. The latter, therefore, do not perform complex tasks, but simple activities of signal reception, processing and transmission to the other neurons. In MLPs a neuron belonging to a given layer is devoted to:

- receiving as input the activation of the neurons belonging to the lower layer, modulated by the value of the connecting weights;

- summing this input from all of the neurons of the lower layer and processing it through a simple non-linear function (the *activation function*);

- transmitting the resulting activation value, multiplied by the connecting weights to all the units of the upper layer.

A difference can be made between the units according to their position in the MLP. Input layer neurons can be considered as simple nodes that transmit the input signals as they are. Therefore they are characterized by a linear activation function. Output layer units can embed different types of activation functions, but there is no theoretical reason to implement a particular type; therefore in the present discussion linear activation will be adopted.

The choice of the activation function embedded in the hidden neurons is a key question, as will be outlined later on in the book. Indeed, the MLP structure's capability is determined by this choice. It can be said first that an MLP with linear hidden units does not offer anything more than a single layer network, as regards approximation capabilities, and therefore only linearly separable functions can be suitably interpolated.

Since the aim of hidden units is to create an alternative representation of the problem, with respect to the input one, a non-linear activation function is needed.

Among the non-linear activation functions there is a certain degree of choice, and the type of learning algorithm can guide the final selection. If the gradient descent approach is used, it is necessary to employ differentiable activation functions. Another key question affecting the final choice of the activation function is the metric adopted for the approximation accuracy. If a uniform approximation is desired, the *sigmoidal activation function* can be employed. In such a way the MLP performs a linear combination of sigmoidal functions which, as outlined above, are universal approximators of continuous, real valued functions.

As has been pointed out in this brief discussion, an MLP is uniquely defined by the following parameters: the number of layers, the number of neurons in each layer and the type of activation function for each layer and neuron. The weight matrix represents the *cognitive aspect*; in fact the problem representation, acquired during the learning phase, and also its degree of accuracy are embedded in the connection weights.

1.2   The learning algorithm

A common characteristic of neural processing is the need for a training phase, in which a *rule* is learned to achieve the desired mapping of the input data so as to match the desired target at the network output. This rule should also be able to perform the same mapping when new input data, not encountered during the learning phase, is presented to the network input. This characteristic is also called the generalization property, and consists of extracting from the examples presented during the learning phase a rule which will also be valid in the new cases. This is perhaps the most common property that living and artificial neural systems share. The learning strategy takes place through a suitable modulation of the network weights. For neural networks two main types of learning strategies are available:

- supervised learning: a kind of *teacher* is present, which shows, for any input pattern, the corresponding target to be achieved at the end of the learning phase. This process is achieved by modulating the connection weights to reach a global configuration in which the (suitably defined) *distance* between any learning pattern and the corresponding target is minimized;

- unsupervised learning: no information is available to guide the learning process for each pattern. The neural network has to *cluster* the input patterns on the basis of the *similarity* between them. Clusters are defined during the learning phase so that all patterns belonging to a given cluster are more similar to one another than to any other pattern belonging to any other cluster.

For MLPs a powerful supervised learning algorithm, *Back Propagation*, is available. Below, the main steps of the algorithm are presented, only to serve as a reference for the following chapters, where some generalizations regarding Back Propagation will be introduced. Only MLPs with one hidden layer will be considered, since they are able to reach the same approximation capabilities as MLPs comprising an arbitrary number of hidden layers.
Once the network topology has been fixed, the network weights are randomly selected.

- *Feedforward phase:* a pattern is presented to the network input.
  Each input unit transmits its output, which, multiplied by the weight

connecting the unit to the units belonging to the hidden layer, forms the
input to this layer.
Each hidden unit performs a summation of the signals coming from the
input layer, processes it by means of its activation function and produces
an output signal.
The weights connecting the hidden layer neurons to the output layer neurons modulate the hidden unit outputs, providing an input signal to the
output units.
Each output layer simply sums the signals coming from the hidden units
directly connected to it and provides an output signal. The signal coming
from the output layer units constitutes the network output.

- *Back Propagation phase* . A comparison takes place between the network
  output and the target for the input pattern presented, based on a defined
  distance, for instance the LMS criterion. Consequently an error signal
  is generated, which is propagated backward through the network layers,
  and the weights are modulated so as to reduce this error (*generalized delta
  rule*), [13]. The *gradient descent* strategy is adopted, i.e. each weight is
  updated by a quantity proportional to the opposite gradient of the error
  with respect to the weight itself.

More precisely, if w_{ij} represents the weight connecting the $i-th$ unit of a given
layer to the $j-th$ unit of the lower layer, o_i and o_j are the outputs of these
units, and E_p represents the error of the pattern p, we have:

$$o_{pi} = f_i(net_{pi}) = f_i(\sum_j w_{ij}o_{pj}) \qquad (1.1)$$

$$E_p = \frac{1}{2}\sum_i (t_{pi} - o_{pi})^2 \qquad (1.2)$$

$$\delta w_{ij} = -k\frac{\partial E_p}{\partial w_{ij}} \qquad (1.3)$$

in which the index i runs over the output layer units, and t_{pi} and o_{pi} respectively
represent the target and the network output for unit i when the pattern p is
presented. Moreover, δw_{ij} represents the weight variation according to the
gradient descent technique. The following equations also hold:

$$\frac{\partial E_p}{\partial w_{ij}} = \frac{\partial E_p}{\partial net_{pi}}\frac{\partial net_{pi}}{\partial w_{ij}} \qquad (1.4)$$

$$\frac{\partial net_{pi}}{w_{ij}} = o_{pj}. \qquad (1.5)$$

If we denote:

$$\delta_{pi} = -\frac{\partial E_p}{net_{pi}} \tag{1.6}$$

since

$$\delta_{pi} = -\frac{\partial E_p}{\partial net_{pi}} = -\frac{\partial E_p}{\partial o_{pi}}\frac{\partial o_{pi}}{\partial net_{pi}} \tag{1.7}$$

in which

$$\frac{\partial o_{pi}}{\partial net_{pi}} = f_i'(net_{pi}) \tag{1.8}$$

the first term in equation (1.7) can be calculated when o represents the output of a neuron belonging either to the output layer or to the hidden one. In the first case we get:

$$\frac{\partial E_p}{\partial o_{pi}} = (t_{pi} - o_{pi}) \tag{1.9}$$

and relation (1.7) becomes:

$$\delta_{pi} = (t_{pi} - o_{pi})f_i'(net_{pi}) \tag{1.10}$$

In the second case we have:

$$\frac{\partial E_p}{\partial o_{pi}} = \sum_k \frac{\partial E_p}{\partial net_{pk}}\frac{\partial net_{pk}}{\partial o_{pi}} = -\sum_k \delta_{pk} w_{ki} \tag{1.11}$$

Here the index k runs over the units belonging to the upper layer connected to the unit o_i, and equation (1.7) gives:

$$\delta_{pi} = f_i'(net_{pi}) \sum_k \delta_{pk} w_{ki}. \tag{1.12}$$

The δ term for a hidden unit, for which no target signal is specified, is therefore determined by means of the δ terms calculated for the units belonging to the upper layer. The output error derivative is back propagated recursively through the network layers to reach a final configuration in which the error with respect to the whole set of learning patterns has been minimized. This is why the algorithm is called *Back Propagation* [13]. This algorithm can be only applied if differentiable activation functions are employed, such as the classical sigmoid:

$$f(o_{pi}) = \frac{1}{1 + e^{-(\sum_j w_{ij} o_{pj} + b_j)}} \tag{1.13}$$

in which the b_j (*bias*) term can be considered as a weight coming from a unit with a unit activation value. Formally the bias term is fundamental because it allows translations to the sigmoid.

In brief, the Back Propagation algorithm involves recursive updating of a generic weight of the following type:

$$\Delta w(n) = -\epsilon \frac{\partial E_p}{\partial w}(n), \qquad (1.14)$$

where n represents the learning cycle index and ϵ is the *learning rate*, which modulates the absolute value of the weight variation in the direction fixed by the error gradient. Too small a value could make the learning process time-consuming, while in the opposite case oscillations could occur. For this reason a *momentum* term has been introduced, so as to take the weight variation (and therefore the error gradient direction) at the preceding learning step into consideration as well. The following equation holds:

$$\Delta w(n) = -\epsilon \frac{\partial E_p}{\partial w}(n) + \alpha \Delta w(n-1), \qquad (1.15)$$

and thus the updated weight will be:

$$w(n+1) = w(n) + \Delta w(n). \qquad (1.16)$$

Chapter 2

Neural networks in complex algebra

2.1 Introduction

In Chapter 1 some fundamental results on the approximation capabilities of classical MLPs in the set of continuous, real valued functions, were given. This chapter extends those results to complex valued functions. The effort is fully justified by the fact that growing interest has been shown in finding suitable devices to efficiently approximate complex valued signals in several scientific areas, such as electromagnetic fields, telecommunications [4], [5] and optical phenomena. In this chapter some conditions on the approximation capabilities of multi-layer neural networks with complex valued weights are formulated and proven. The first result states that a simple extension to the complex field of the real valued sigmoidal function has limited approximation performance. A further theorem states the universal interpolation properties of complex valued functions for sequences of a particular type of bounded, not analytic, complex valued sigmoidal function, by using some characteristics of real hyperplanes in complex subspaces and employing the $Hahn - Banach$ and $Rietz$ representation theorems in the complex field.

2.2 The Complex Multi-Layer Perceptron

In the world of neural network research, novel structures are often defined, which differ from existing ones in the type of connectivity between neurons, the neuron activation function, the type of weight adaptation law, or the various strategies introduced to improve network performance. Each of them is suitable to manage and solve a particular class of problems. Since 1990, several neural structures based on the Multi-Layer Perceptron have appeared in literature, and are able to deal with problems of interpolation and identification of systems defined in complex algebra; for such networks the acronym CMLP (Complex Multi-Layer Perceptrons) has therefore been devised [10], [11], [12], [2], [3].

The structure of a CMLP is equivalent to that of a real classical MLP in which the input and output signals, the weights and the bias values, are no longer real quantities, but complex numbers, and the activation functions embedded in the neurons belonging to the hidden layers are complex valued functions.

Several scientific works have recently appeared in literature, showing CMLP applications in electromagnetic fields and telecommunications. Indeed, although a complex variable can be treated by a real MLP as a pair of real variables, the introduction of a new neural structure, able to manage complex variables directly, presents some clear advantages. In fact, in order to process a function of n complex variables by means of a real MLP, $2n$ input units are needed and, when the function to be approximated assumes its codomain in C^m, $2m$ output units are needed. The number of connections that link the network with the external world is therefore considerably high, apart from the required number of hidden units. The high number of degrees of freedom therefore reduces the efficiency of the learning algorithm, slowing down the convergence rate and affecting network performance, with a high probability of getting stuck in local minima, as will be made clearer in the following sections. The use of a neural structure which is able to process complex signals directly with only one input unit for each signal makes it possible to reduce the number of input and output units of the network by a half, thus reducing the number of connections but maintaining the same number of interpolating functions; the performance of the learning algorithm is therefore improved. Moreover, the feedforward phase in the CMLP requires a lower number of multiplications than the real equivalent MLP, thus decreasing the computational complexity of the network [5].

The various CMLP structures introduced in literature differ according to the type of activation function embedded in the neurons. In particular, the CMLP proposed by Kim [10] and by Leung [11] uses an analytic activation function, while the one proposed by Piazza [12], Nitta [2] and Koutsougeras [3] implements a not analytic activation function, as described below. The choice of the activation function heavily characterizes the approximation capabilities of the neural network.

2.3   Notations

The notations used to describe the structure and the learning algorithm for the CMLP neural networks introduced above are:

- i=$\sqrt{-1}$;

- M: number of layers in the network;

- l: layer index (in particular $l = 0$ denotes the input layer and $l = M$ the output one);

- N_l: number of neurons in the l-th layer;

- $\mathbf{S}_n^l = S_{0n}^l + iS_{1n}^l \in C$:'net' function relative to the n-th neuron of layer l;

- $\mathbf{X}_n^l = X_{0n}^l + iX_{1n}^l \in C$: output of the n-th neuron in the l-th layer (in particular $\mathbf{X}_0^l = 1 + i$ represents the bias input for the l-th layer, $\mathbf{X}_n^0 = i_n$ where $n = 1, \ldots, N_0$ is the input to the network and $\mathbf{X}_n^M = \mathbf{Y}_n$ where $n = 1, \ldots, N_M$ is the output);

- $\mathbf{w}_{nm}^l = w_{0nm}^l + iw_{1nm}^l \in C$: connection weight between the n-th neuron of the l-th layer and the m-th one of the $l - 1$-th layer;

- $\sigma_n^l(x + iy) \ C \rightarrow C$: activation function of the n-th neuron of the l-th layer;

- $\dot{\sigma}_n^l(x + iy)$: activation function derivative of the n-th neuron of layer l;

- $\mathbf{t}_n = t_{0n} + it_{1n}, \ n = 1, \ldots, M$: target for the n-th output.

The error function for each pattern is defined as:

$$E = \frac{1}{2} \sum_{i=1}^{N_M} (\mathbf{t}_i - \mathbf{Y}_i)(\mathbf{t}_i - \mathbf{Y}_i)^* \tag{2.1}$$

where '*' denotes the conjugation operator.

2.4 CMLP with an analytic activation function

The Multi-Layer Perceptron described in this section [11], implements an activation function for the hidden layer neurons which corresponds to the natural generalization to the complex space of the real valued sigmoidal function:

$$f(x) = \frac{1}{1 + exp(-x)} \tag{2.2}$$

Therefore the complex valued sigmoidal function for each neuron is the following:

$$\sigma_n(z) = \frac{1}{1 + exp(-z)} \tag{2.3}$$

where $z = x + iy$ is the complex variable.
It should be pointed out that $\sigma(z)$ is analytic and unbounded in C; in particular

it has infinite singularities for $z \in 0 \pm i(n+1)\pi$.

The same activation function was proposed by Kim [10], who derived a learning algorithm for CMLPs with only one hidden layer. In the following a Multi-Layer Perceptron with M layers will be considered.

The equations describing the processing during the 'forward' phase are:
with $l = 1, \ldots, M$ and $n = 1, \ldots, N_l$

$$\mathbf{S}_n^l = \sum_{m=0}^{N_{l-1}} \mathbf{w}_{nm}^l \mathbf{X}_m^{l-1} \tag{2.4}$$

$$\mathbf{X}_n^l = \sigma(\mathbf{S}_n^l) \tag{2.5}$$

The learning algorithm for the proposed network was derived in [11] in the same way as the classical Back Propagation algorithm (BP) [13], by minimizing the error function with respect to each weight with a steepest gradient descent technique and applying the chain rule to obtain the partial derivatives of the error function with respect to the weights belonging to the hidden layers.

Since the activation function is analytic, the first derivative exists and its partial derivatives with respect to the real and imaginary parts satisfy the Cauchy-Riemann conditions, namely:

$$u_x = v_y \tag{2.6}$$

$$u_y = -v_x \tag{2.7}$$

$$\dot{f} = u_x + iv_x \tag{2.8}$$

where:

$$f(x+iy) = u(x,y) + iv(x,y) \tag{2.9}$$

$$u_x = \frac{\partial u}{\partial x} \quad u_y = \frac{\partial u}{\partial y} \tag{2.10}$$

$$v_x - \frac{\partial v}{\partial x} \quad v_y = \frac{\partial v}{\partial v} \tag{2.11}$$

Moreover, since the activation function does not contain complex operations, we have:

$$\mathbf{X}_n^* = f(\mathbf{S}_n^*) \tag{2.12}$$

and therefore:

$$\frac{\partial \mathbf{X}_n}{\partial \mathbf{S}_n^*} = \frac{\partial \mathbf{X}_n^*}{\partial \mathbf{S}_n} = 0 \tag{2.13}$$

The following relations are derived:

with $m = 0, \ldots, N_{l-1}$ and with $n = 1, \ldots, N_l$

$$\Delta \mathbf{w}_{nm}^l = \epsilon \delta_n^l \mathbf{X}_m^{*l-1} \tag{2.14}$$

where:

$$\delta_n^l = (\mathbf{t}_n - \mathbf{Y}_n)\dot{\sigma}(\mathbf{S}_n^{*l}) \tag{2.15}$$

for the units belonging to the output layer and:

$$\delta_n^l = \dot{\sigma}(\mathbf{S}_n^{*l}) \sum_{k=1}^{N_{l+1}} \mathbf{w}_{kn}^{*l+1} \delta_k^{l+1} \tag{2.16}$$

for the units belonging to the hidden layers.
$\dot{\sigma}$ represents the activation function derivative, given by:

$$\dot{\sigma}(z) = \sigma(z)(1 - \sigma(z)) \tag{2.17}$$

During the implementation of the learning algorithm it is necessary to take into consideration that, owing to the singularities of the activation function, the weights have to be confined in suitable regions [11]. Such a solution, besides being difficult to achieve in an algorithm like the Back Propagation one, which does not consider any weight control mechanism, could prevent the error function from reaching the global minimum. The same considerations also hold when other activation functions commonly employed for the real MLP are used in extension to the complex space: for example the function $f(z) = tanh(z) \rightarrow \infty$ for $z \in 0 \pm i(\frac{2n+1}{2})\pi$. Moreover a greater drawback lies in the reduced approximation capabilities of such a network, as will be discussed in a following section.

2.5 CMLP with a not analytic activation function

Besides the CMLP described in the previous section, another neural structure was proposed in literature in roughly the same period. This structure, of the CMLP type, implements a sigmoidal activation function which is not analytic and bounded; more specifically, in [12] a learning algorithm was introduced for a network with M hidden layers while in [2] a network with only one hidden layer was considered. In both cases the activation function chosen is the following:

$$\sigma(x + iy) = \frac{1}{1 + exp(-x)} + i\frac{1}{1 + exp(-y)} \tag{2.18}$$

This function is not analytic and bounded in C. The absence of the first derivative of this activation function does not prevent the determination of a learning algorithm for the CMLP; in fact, in order to derive weight updating formulas, the following conditions have to be satisfied:

- there exist the partial derivatives u_x, u_y, v_x, v_y;

- the partial derivatives are bounded;

- $u_x v_y \neq v_x u_y$.

The first condition is needed in order to calculate the error function gradient with respect to each weight and the second one derives from the fact that the partial derivatives are present in the formulas of the weight updating. If the last condition is not met, the weights cannot be updated even when a non-vanishing error between the target and the output exists [3].

The activation function that will be used in the following, i.e. function (2.18) satisfies all the required conditions.

The forward phase of the network proposed is described by the formulas: for $l = 1, \ldots, M$ and $n = 1, \ldots, N_l$

$$\mathbf{S}_n^l = \sum_{m=0}^{N_{l-1}} \mathbf{w}_{nm}^l \mathbf{X}_m^{l-1} \tag{2.19}$$

$$\mathbf{X}_n^l = \sigma(\mathbf{S}_n^l) \tag{2.20}$$

The learning algorithm derived in [12] is reported in the following: for $m = 0, \ldots, N_{l-1}$ and for $n = 1, \ldots, N_l$:

$$\Delta \mathbf{w}_{nm}^l = \epsilon \delta_n^l \mathbf{X}_m^{*l-1} \tag{2.21}$$

where:

$$\delta_n^l = (\mathbf{e}_n^l) \odot \dot{\sigma}(\mathbf{S}_n^l) \tag{2.22}$$

and:

$$\mathbf{e}_n^l = \begin{cases} \mathbf{t}_n - \mathbf{Y}_n & \text{for } l = M \\ \sum_{k=1}^{N_{l+1}} \mathbf{w}_{ik}^{*l+1} \delta_k^{l+1} & \text{for } l = M - 1, \ldots, 1 \end{cases} \tag{2.23}$$

for the units belonging to the hidden layers.

The notation '$\odot$' denotes the component by component product between two complex numbers, therefore we get.

$$\mathbf{z} \odot \mathbf{z}' = (x + iy) \odot (x' + iy') = xx' + iyy'$$

The derivative is computed as:

$$\dot{\sigma}(x + iy) = \dot{\sigma}(x) + i\dot{\sigma}(y) \tag{2.24}$$

It should be observed that the learning algorithm for the CMLP becomes a real Back Propagation one if the input signals are real-valued. Moreover, if the neural network contains only one hidden layer with a linear activation function, the algorithm reduces to the complex LMS algorithm [14].

Another activation function, complex-valued, not analytic and bounded, which obeys the conditions given above, has been proposed in literature [3], namely:

$$\sigma(z) = \frac{z}{c + \frac{1}{r}|z|} \quad (2.25)$$

where the parameter c modulates the slope and r limits the range of function values. The function (2.25) maintains the phase of its argument unvaried.

The feedforward phase is described by the previous equations, replacing the activation function with (2.25). The weight updating formulas, here again derived with the gradient technique, are the following:

for $m = 0, \ldots, N_{l-1}$

$$\Delta \mathbf{w}_{nm}^l = \epsilon \delta_n^l \mathbf{X}_m^{*l-1} \quad (2.26)$$

where:

$$\delta_n^M = \Re(\mathbf{t}_n - \mathbf{Y}_n)(u_x(\mathbf{S}_n^M) + iu_y(\mathbf{S}_n^M)) + \Im(\mathbf{t}_n - \mathbf{Y}_n)(v_x(\mathbf{S}_n^M) + iv_y(\mathbf{S}_n^M)) \quad (2.27)$$

for the units of the output layer and:

$$\delta_n^l = (u_x(\mathbf{S}_n^l) + iu_y(\mathbf{S}_n^l))\Re\left(\sum_{k=1}^{N_{l+1}} \mathbf{w}_{ik}^{*l+1} \delta_k^{l+1}\right) + (v_x(\mathbf{S}_n^l) + iv_y(\mathbf{S}_n^l))\Im\left(\sum_{k=1}^{N_{l+1}} \mathbf{w}_{ik}^{*l+1} \delta_k^{l+1}\right) \quad (2.28)$$

for the hidden layer units, where:

$$u_x = \begin{cases} \dfrac{r(y^2 + cr|z|)}{|z|(cr + |z|^2)} & \text{if } |z| \neq 0 \\ \dfrac{1}{c} & \text{if } |z| = 0 \end{cases} \quad (2.29)$$

$$u_y = \begin{cases} -\dfrac{rxy}{|z|(cr + |z|^2)} & \text{if } |z| \neq 0 \\ 0 & \text{if } |z| = 0 \end{cases} \quad (2.30)$$

$$v_x = \begin{cases} -\dfrac{rxy}{|z|(cr + |z|^2)} & \text{if } |z| \neq 0 \\ 0 & \text{if } |z| = 0 \end{cases} \quad (2.31)$$

$$u_x = \begin{cases} \dfrac{r(x^2 + cr|z|)}{|z|(cr + |z|^2)} & \text{if } |z| \neq 0 \\ \dfrac{1}{c} & \text{if } |z| = 0 \end{cases} \quad (2.32)$$

are the activation function partial derivatives.

The formulas given are similar to (2.21), taking into account that in (2.21), owing to the particular form of the activation function, $u_y = v_x = 0$.

2.6   Approximation capabilities of the CMLP

Relevant theoretical results about the approximation capabilities of the real MLP in the set of real-valued functions were independently obtained by various authors in the same period, i.e. around 1989 [15], [19], [20].

One of these, as reported in the first chapter, was obtained by Cybenko [15] and will be extended to complex algebra in this section. The approximation capabilities of CMLPs heavily depend on the type of activation function embedded in the neurons. In fact a number of theorems have been proven on this topic [22], [23]:

Theorem 2.6.1 *Let D be a compact convex domain in C^n, and let ϕ_n be a sequence of analytic complex-valued functions converging uniformly to ϕ on D. Then ϕ is analytic in D^0 (the inner part of D).*

Proof ϕ is continuous because it is the uniform limit of continuous functions. Since the domain D is convex, ϕ is analytic if and only if its restriction to any complex line intersecting D is an analytic function of one complex variable. Actually, it is sufficient to show that the function is holomorphic in each variable separately [7]. Let:

$$D_l = D \bigcap \{\bar{z}_0 + \lambda \bar{z}_1 \mid \lambda \in \mathcal{C}, \bar{z}_i \in C^n, i = 0, 1\} \tag{2.33}$$

be such an intersection and let us say $\phi_l(\lambda) = \phi\{\bar{z}_0 + \lambda \bar{z}_1\}$ is the restriction of ϕ to D_l. By Morera's theorem [8], ϕ_l is analytic if and only if:

$$\int_{\partial \Delta} \phi_l(z) dz = 0$$

for any closed triangle $\Delta \subset D_l$. This condition is assured by the analytic nature of $\phi_{n,l}$:

$$\int_{\partial \Delta} \phi_{n,l}(z) dz = 0 \quad \forall n. \tag{2.34}$$

Going to the limit under the integral, by uniform convergence, it follows that:

$$\int_{\partial \Delta} \phi(z) dz = 0. \tag{2.35}$$

The theorem which states the approximation capabilities of the CMLP is the following:

Theorem 2.6.2 *Linear combinations of the type:*

$$G(z) = \sum_{i=1}^{N} \alpha_i \sigma(w_i^t z + \theta)]$$

where σ is an activation function of type (2.3), are not dense in $C^0(D_n)$.

Proof
It is sufficient to prove the theorem in the case $G(z) : D \to C$, its generalization to $G(z) : D^n \to C^m$ being straightforward. Let us consider a function $G(z) : D \to C$, where $D = \{z \in C :| z | \leq 1\}$, such that its maximum modulus is reached in an inner point of D. Such a function cannot be analytic due to the maximum modulus theorem. Therefore, as a consequence of theorem 2.6.1, $G(z)$ is not the limit of a sequence of activation functions of the type (2.3).

Remarks
As a consequence of these results it can easily be derived that not analytic function approximation cannot be performed by employing superpositions of analytic functions. In the complex case the condition of not analyticity is, in fact, more restrictive than in the real one, in which, taking into consideration the Stone-Weierstrass theorem, real analytic functions (e.g. polynomials) can approximate continuous but not analytic functions. Taking into account what was said previously, it would be desirable to find a function able to approximate a wider class of complex valued functions. The theoretical analysis reported below states that linear combinations of type (2.18) can approximate any continuous complex valued function defined in the unit polydisk. In particular, Theorem 2.6.3, performing a suitable partition of the complex domain, shows that functions of type (2.18) are discriminatory. In the proof of Theorem 2.6.4 the results derived in theorem 2.6.3, together with the Hahn-Banach and Rietz-Representation Theorems in the complex case, are used to obtain a density condition in $C^0(D_n)$ for the complex subspace generated by the complex function (2.18).

Theorem 2.6.3 *The function*

$$f(z) = \frac{1}{1 + exp(-x)} + i\frac{1}{1 + exp(-y)}$$

is discriminatory.

Proof Let $\mu(z)$ be a complex finite regular Borel measure on D_n such that:

$$\int_{D_n} f(\bar{w}^t \bar{z} + \theta) d\mu(z) = 0 \quad \forall \bar{w}, \bar{z} \in C^n, \forall \theta \in C.$$

Having fixed $\lambda \in \mathcal{R}$, let us consider the following function:

$$\sigma[(\lambda \Re(\bar{w}^t \bar{z} + \theta) + b], \quad \forall b \in \mathcal{R}, \forall \bar{w} \neq 0,$$

where $\Re(z)$ denotes the real part of a complex number z, and let us evaluate the following pointwise limit:

$$\lim_{\lambda \to \infty} \sigma[(\lambda \Re(\bar{w}^t \bar{z} + \theta) + b] = \Psi_R[(\bar{w}^t \bar{z} + \theta) + b].$$

The result is:

$$\Psi_R[(\bar{w}^t\bar{z} + \theta) + b] = \begin{cases} 1 & \text{if } \Re(\bar{w}^t\bar{z} + \theta) > 0 \\ 0 & \text{if } \Re(\bar{w}^t\bar{z} + \theta) < 0 \\ \sigma(b)\epsilon]0,1[& \text{if } \Re(\bar{w}^t\bar{z} + \theta) = 0 \end{cases} \qquad (2.36)$$

Similarly, $\forall\, c \in R, \forall\, \bar{w} \neq 0$ let us consider:

$$\sigma[(\lambda\Im(\bar{w}^t\bar{z} + \theta) + b]\quad and$$

$$\int_{D_n} f(\bar{w}^t z + \theta)d\mu(z) = 0 \quad \forall \bar{w} \in C^n, \forall \theta \in C$$

in which $\Im(z)$ denotes the imaginary part of the complex number z and:

$$\Psi_I[(\bar{w}^t\bar{z} + \theta) + c] = \begin{cases} 1 & \text{if } \Im(\bar{w}^t\bar{z} + \theta) > 0 \\ 0 & \text{if } \Im(\bar{w}^t\bar{z} + \theta) < 0 \\ \sigma(c)\epsilon]0,1[& \text{if } \Im(\bar{w}^t\bar{z} + \theta) = 0 \end{cases} \qquad (2.37)$$

By the *Lebesgue-dominated Convergence Theorem* [9] it follows that:

$$\int_{D_n} \Psi d\mu(z) = \int_{D_n} (\Psi_R + i\Psi_I)d\mu(z) =$$

$$= \lim_{\lambda\to\infty} \int_{D_n} \sigma\{[(\lambda\Re(\bar{w}^t\bar{z} + \theta) + b] + i[(\lambda\Im(\bar{w}^t\bar{z} + \theta) + c]\}d\mu(z) = 0 \quad (2.38)$$

Now, for $\bar{w} \neq \bar{0}$ let us call:

$$H_{++} = \{\bar{z} : \Re[(\bar{w}^t\bar{z} + \theta) + b] > 0\}\bigcap\{\bar{z} : \Im[(\bar{w}^t\bar{z} + \theta) + c] > 0\}$$

$$H_{+-} = \{\bar{z} : \Re[(\bar{w}^t\bar{z} + \theta) + b] > 0\}\bigcap\{\bar{z} : \Im[(\bar{w}^t\bar{z} + \theta) + c] < 0\}$$

$$H_{-+} = \{\bar{z} : \Re[(\bar{w}^t\bar{z} + \theta) + b] < 0\}\bigcap\{\bar{z} : \Im[(w^t\bar{z} + \theta) + c] > 0\}$$

$$H_{--} = \{\bar{z} : \Re[(\bar{w}^t\bar{z} + \theta) + b] < 0\}\bigcap\{\bar{z} : \Im[(w^t\bar{z} + \theta) + c] < 0\}$$

$$H_{0+} = \{\bar{z} : \Re[(\bar{w}^t\bar{z} + \theta) + b] = 0\}\bigcap\{\bar{z} : \Im[(w^t\bar{z} + \theta) + c] > 0\}$$

$$H_{0-} = \{\bar{z} : \Re[(\bar{w}^t\bar{z} + \theta) + b] = 0\}\bigcap\{\bar{z} : \Im[(w^t\bar{z} + \theta) + c] < 0\}$$

$$H_{+0} = \{\bar{z} : \Re[(\bar{w}^t\bar{z} + \theta) + b] > 0\}\bigcap\{\bar{z} : \Im[(w^t\bar{z} + \theta) + c] = 0\}$$

$$H_{-0} = \{\bar{z} : \Re[(\bar{w}^t\bar{z} + \theta) + b] < 0\}\bigcap\{\bar{z} : \Im[(w^t\bar{z} + \theta) + c] = 0\}$$

$$H_{00} = \{\bar{z} : \Re[(\bar{w}^t\bar{z} + \theta) + b] = 0\}\bigcap\{\bar{z} : \Im[(w^t\bar{z} + \theta) + c] = 0\}$$

Taking into account (2.36) and (2.37), let us evaluate the following integral:

$$\int_{C^n} \Psi_R[(\bar{w}^t\bar{z} + \theta) + b]d\mu(z) + i\int_{C^n} \Psi_I[(\bar{w}^t\bar{z} + \theta) + c]d\mu(z) =$$

$$= (1+i)\mu(H_{++}) + \mu(H_{+-}) + i\mu(H_{-+}) +$$

$$+[\sigma(b) + i]\mu(H_{0+}) + \sigma(b)\mu(H_{0-}) + [i\sigma(c) + 1]\mu(H_{+0}) +$$

$$+i\sigma(c)\mu(H_{-0}) + [\sigma(b) + i\sigma(c)]\mu(H_{00}) = 0.$$

In order to make the following notation simpler, let us group as k all the quantities not depending on $\sigma(b)$ and on $\sigma(c)$. We get:

$$\begin{aligned}
0 &= K + \sigma(b)\mu(H_{0+}) + \sigma(b)\mu(H_{0-}) + i\sigma(c)\mu(H_{+0}) + \\
&\quad +i\sigma(c)\mu(H_{-0}) + \sigma(b)\mu(H_{00}) + i\sigma(c)\mu(H_{00}) = \\
&= K + \sigma(b)[\mu(H_{0+}) + \mu(H_{0-}) + \mu(H_{00})] + \\
&\quad +i\sigma(c)[\mu(H_{+0}) + \mu(H_{-0}) + \mu(H_{00})] = \\
&= K + \sigma(b)\mu(H_\star) + i\sigma(c)\mu(H_{\star\star}) = 0 \qquad (2.39)
\end{aligned}$$

in which:

$$H_\star = \{\bar{z} : \Re(\bar{w}^t\bar{z} + \theta) = 0\}$$

$$H_{\star\star} = \{\bar{z} : \Im(\bar{w}^t\bar{z} + \theta) = 0\}.$$

Relation (2.39) holds $\forall\sigma(b)\epsilon]0,1[$ and $\forall\sigma(c)\epsilon]0,1[$.
In this case it can easily be shown (see lemma 2.6.1) that:

$$\mu(H_\star) = \mu(H_{\star\star}) = 0 \qquad (2.40)$$

From expression 2.39 it can therefore be seen that $K = 0$, which is:

$$K = (1+i)\mu(H_{++}) + \mu(H_{+-}) + i\mu(H_{-+}) + i\mu(H_{0+}) + \mu(H_{+0}) = 0$$

Grouping $\Re$ and $\Im$ we get:

$$\mu(H_{++}) + \mu(H_{+-}) + \mu(H_{+0}) + i[\mu(H_{++}) + \mu(H_{-+}) + \mu(H_{0+})] = 0$$

that is:

$$\mu(\Re(\bar{w}^t\bar{z} + \theta) > 0) + i\mu(\Im(\bar{w}^t\bar{z} + \theta) > 0) = 0.$$

Now, let $\theta = -a - ib$: it follows that:

$$\mu(\Re(\bar{w}^t\bar{z}) > a) = -i\mu(\Im(\bar{w}^t\bar{z}) > b), \quad \forall\bar{w}\epsilon C^n, \ \forall a, b, \epsilon\mathcal{R}.$$

Now, keeping b fixed and letting a vary, it follows that:

$$\mu(\Re(\bar{w}^t\bar{z}) > a) = \mu(\Re(\bar{w}^t\bar{z}) > a') \quad \forall a < a', \ a'\epsilon\mathcal{R}$$

and therefore:

$$\mu(a < \{\Re(\bar{w}^t \bar{z})\} \le a') = 0.$$

But, letting a' go to infinity, on account of the σ-properties of measures we get:

$$\mu(\Re(\bar{w}^t \bar{z}) > a) = 0, \quad \forall a \epsilon \mathcal{R}.$$

It can be observed that any open real semispace S of C^n can be written in the form:

$$S = \{z : \Re(\bar{w}^t \bar{z} + \theta) > 0\}.$$

Therefore, for any real open semispace in C^n :

$$\mu(S) = \Re(\mu(S)) + i\Im(\mu(S)) = 0.$$

Now, owing to the fact that we are analysing real measures and real semispaces, referring directly to the Cybenko proof of Lemma 1 [15] we can show that:

$$\Re(\mu(S)) = \Im(\mu(S)) = 0,$$

implies that $\mu(z) = 0$.

Lemma 2.6.1 *Let $K, z, w \in C$ be fixed, and let us assume that:*

$$k + az + i(bw) = 0 \quad \forall a, b \epsilon]0, 1[\tag{2.41}$$

Then:

$$z = w = 0.$$

Proof
Let us assume that $z \ne 0$. From (2.41) we get:

$$a = \frac{-(K + i(bw))}{z}. \tag{2.42}$$

Fixing a arbitrarily, relation 2.42 holds $\forall\ b\ \in\ ]0,\ 1[$, but this is clearly impossible, K, z e w being fixed by the hypothesis. The same considerations can be derived in the case $w \ne 0$.

Theorem 2.6.4 *Linear combinations of functions of the type (2.18) are dense in $C^0(D_n)$.*

Proof
Let S be the complex linear subspace of $C^0(D_n)$ generated by the function:

$$\sigma(\bar{w}^t \bar{z} + \theta) = \sigma_r(\Re(\bar{w}^t \bar{z} + \theta)) + i\sigma_r(\Im(\bar{w}^t \bar{z} + \theta)) \tag{2.43}$$

with $\bar{w} \in C^n$, $\theta \in C$, e σ_r a fixed real sigmoidal type function. Let us assume that S is not dense in $C^0(D_n)$, i.e. the closure of S, say R is not all of $C^0(D_n)$.

R is a closed proper subspace of $C^0(D_n)$; then, by the *complex Hahn-Banach theorem* [8], there is a bounded linear functional $L : C^0(D_n) \to C$, such that $L(R) = L(S) = 0$, but $L \neq 0$. In particular, since $\forall \bar{w}$ e θ, $\sigma(\bar{w}^t \bar{z} + \theta) \epsilon \Re$, we get: $L(\sigma(\bar{w}^t \bar{z} + \theta)) = 0$. By the *Rietz representation theorem in the complex case* [8], for any bounded linear functional L on $C^0(D_n)$, there exists a complex Borel regular measure $\mu(z)$ on $C^0(D_n)$ such that, for each $f \in C^0(D_n)$, we can write:

$$L(f) = \int_{D_n} f d\mu(z).$$

In particular, taking as f the function expressed in (2.43) we get:

$$L(\sigma(\bar{w}^t \bar{z} + \theta)) = \int_{D_n} \sigma(\bar{w}^t \bar{z} + \theta) d\mu(z) = 0 \qquad (2.44)$$

$$\forall \bar{w} \in C^n, \ \ \forall \theta \in C$$

Since σ is discriminatory, it follows that $\mu(z) = 0$, contradicting the assumption $L \neq 0$. Therefore the subspace S is dense in $C^0(D_n)$.

Such results therefore state that CMLPs with complex valued not analytic activation functions of type (2.18) are universal interpolators of complex valued continuous functions on a compact subset of C^n. Moreover, it has been proven that the CMLP structure described in section 2.4 is only able to approximate analytic functions. No theoretical results exist about the approximation capabilities of this type of CMLP in the analytic function space; therefore it could be possible to approximate all of the analytic functions in a subset of C^n.

More interesting results have, on the other hand, been derived for the CMLP structure with a not analytic and bounded activation function, as described in section 2.5.

2.7 Multi-Layer perceptrons in real or complex algebra: a comparison

In this section a comparison is made between the complex structure with activation function (2.18) and the real MLP as regards the number of parameters employed.

For this purpose, let us consider a CMLP with a not analytic activation function, only one hidden layer, Ni inputs, Nh hidden neurons and No outputs. Let us denote with NCc the number of real weights, i.e. the number of weights obtained considering each complex weight as equal to two real weights; this number is given by:

$$NCc = (Ni * Nh + Nh * No) * 2 \qquad (2.45)$$

The equivalent network, considering the same number of interpolating functions, comprises $2 * Ni$ inputs, $2 * Nh$ hidden neurons and $2 * No$ output units. The number of weights will therefore be:

$$NCr = 2 * Ni * 2 * Nh + 2 * Nh * 2 * No = 2 * NCc \qquad (2.46)$$

Considering the same number of real sigmoidal functions embedded in the network, the CMLP needs half the number of parameters needed by the real MLP (the number of biases being the same in both structures). From experience gained with various applications it has been observed that the number of interpolating functions needed by the CMLP to reach the same performance as the real MLP is slightly greater, but the real parameter saving always remains about $30 - 40\%$.

Another interesting comparison lies in the number of real multiplications performed by the two equivalent structures. Here again it can be derived that the complex MLP requires fewer multiplications to execute the feedforward phase. Each multiplication between complex numbers has been considered equal to four real multiplications and each complex weight equivalent to two real weights.

However, from a merely theoretical point of view, the approximation capabilities of the CMLP with activation function (2.18) and those of the real MLP are the same: the fact that, considering the same number of approximating functions, the complex network requires a lower number of parameters, implies that the learning algorithm has to manage a lower number of parameters. As a consequence, the CMLP has a lower probability of being stuck in local minima configurations, and the generalization performance improves. Moreover, owing to the lower number of parameters to determine, the need for a large training set can, to a certain extent, be relaxed.

2.8   Applications of the CMLP: some examples

The use of CMLP neural networks has proved to be more suitable than the real MLP in all of the applications proposed in literature. For example, the CMLP with an activation function of the type (2.18) was applied in [5] to solve non linear equalization problems. Successful results have been obtained as compared with the real MLP, as regards the error in the learning phase, the computational complexity (time complexity), and the number of parameters used (space complexity).

From the comparisons performed with different topologies it has been observed that, as foreseen, the CMLP is able to reach a smaller error than the real MLP and employs a reduced topology, with a smaller number of parameters and multiplications in the feedforward phase.

A number of examples regarding CMLP approximation capabilities are given in the following section.

2.9 Numerical examples

In this section some numerical examples are given in order to show the advantages deriving from the use of CMLPs in performing function interpolation in the complex domain, both as regards "time complexity" (i.e. the number of products performed during the feedforward phase), and "space complexity" (i.e. the number of real parameters involved in the network). The examples are also intended to further outline what was theoretically proven above. The neural structures considered in this section, as pointed out above, depend on the form of the activation function used in the hidden neurons. For each of these structures, suitable Back-Propagation learning algorithms have been derived as mentioned in the previous sections. The learning algorithms corresponding to each neural structure will be labeled as follows:

- algorithm A: based on the activation function (2.3);

- algorithm B: based on the activation function (2.18);

- algorithm C: classical real BP algorithm.

Without loss of generality, only MLP topologies with one hidden layer and linear activation functions in the output layer will be considered. The notation: $N_{i,h,o}$, indicates the number of input, hidden and output neurons which make up the network topology. For each example several learning phases with different network topologies were carried out for the algorithms being considered: the most significant will be illustrated in tables, labeled with an experiment index.

2.9.1 Interpolation of not analytic functions

Let us consider the following not analytic function:

$$f(x + iy) = e^{iy}(1 - x^2 - y^2) \qquad (2.47)$$

In order to perform the approximation, a set of 500 patterns was randomly generated calculating the function values in the domain interval [0+i0 ,1+i1]. Among these, a set of 200 patterns was used to train the neural architectures, while the remaining 300 were left to perform the testing phase. As regards the number of hidden neurons employed, a growing strategy was used for each type of real and complex structure. In other words, several training phases were performed fixing an equal maximum number of learning cycles. For such topologies a testing phase was also performed in order to evaluate their generalization capabilities on a set of patterns not presented during the learning phase. The learning parameters were fixed for all the experiments to the values:

- learning rate = 0.1;

Algorithm type	A	
Training Number	1	2
Topology	$N_{1,2,1}$	$N_{1,3,1}$
Error	23.59	64.22
No. of parameters	14	26

Algorith type	B		
Training Number	3	4	5
Topology	$N_{1,2,1}$	$N_{1,3,1}$	$N_{1,4,1}$
Error	0.424	0.0723	0.077
No. of parameters	14	20	26

Algorith type	C		
Training Number	6	7	8
Topology	$N_{2,4,2}$	$N_{2,5,2}$	$N_{2,6,2}$
Error	0.126	0.0791	0.1289
No. of parameters	22	27	32

Table 2.1: Function (2.47) approximation

- momentum = 0.01;

- number of cycles = 1100.

The results of this strategy are schematically reported in Table 2.1. From the analysis of these results it can be observed that, as theoretically proven, the CMLP with an analytic activation function (Algorithm A) is not able to perform function approximation for any of the topologies used. In order to better evaluate the results obtained by the other structures (algorithms B and C), the learning error curves are plotted together in Fig.2.1 as a function of the number of cycles for the different cases reported in Table 2.1. In order to make the difference between the error curves corresponding to the various experiments clearer, the error surface only refers to the last part of the learning phases. The figure does not include the training phases of experiments 1 and 2, due to the fact that Algorithm A does not converge at all, as expected from the theoretical results. Such a representation will also be used in all the following examples. In Fig.2.2 the performance of the networks during the testing phase is depicted. For each experiment, labeled in the x-axis, each '*' represents the value of the error norm on each of the 200 patterns employed in the testing phase. The maximum value in the y-axis therefore corresponds to the bound of the error norm for the experiment being considered. The best performance was reached in experiment N. 4, corresponding to the CMLP $N_{1,3,1}$ trained with algorithm B. As regards the real MLP, the best results were obtained in experiment N. 7 which refers to the topology $N_{2,5,2}$. Although the performance

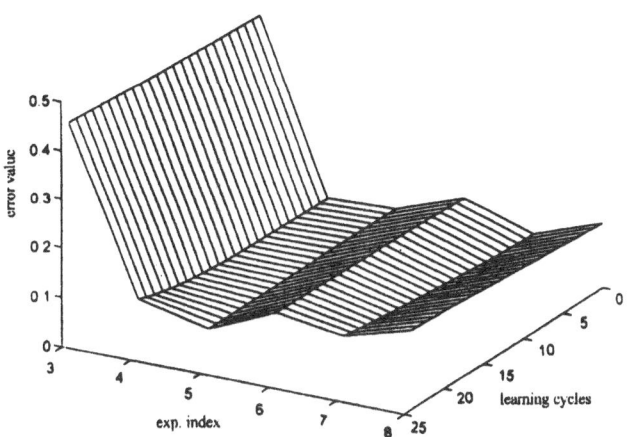

Figure 2.1: *Approximation of function 2.47, learning error trends for the networks of Table 2.1*

levels reached are quite similar, it has to be observed that the number of real parameters involved is very different. From the table it can, in fact, be observed that the CMLP $N_{1,3,1}$ employs 20 real parameters, while the real MLP $N_{2,5,2}$ requires 27 parameters. From these results it can be deduced that even in simple numerical examples the use of CMLPs allows a great saving in real parameters; as a consequence, such a network shows a considerable speed-up in the learning phase and is more likely to reach the global minimum. It can also be observed that the number of multiplications in the feedforward phase is greatly reduced [12].

As a further example let us consider the following not analytic function:

$$f(x + iy) = x^2 + y^2 + 2ixy \qquad (2.48)$$

As in the previous case, in order to approximate this function, different topologies were considered for each of the algorithms. 100 patterns were used to perform the network learning phase, and a set of 100 unknown patterns was used to carry out the testing phase. All of the patterns were determined considering the function values in 200 randomly selected complex points in the interval [0+i0 ,1+i1]. The learning parameters and the number of cycles were fixed to the values used in the previous example. The results obtained are shown in Table 2.2. From an analysis of Table 2.2 it can be observed that here again, as expected from the theoretical results, algorithm A does not converge at all with any topology. The performance of algorithms B and C is quite similar as regards convergence properties, but it can be observed that in order to obtain the same order of magnitude for the training error, algorithm C requires

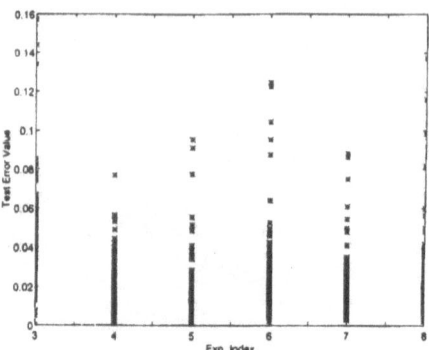

Figure 2.2: *Approximation of function 2.47, testing error absolute values for the networks of Table 2.1*

an MLP with twice the number of neurons for each layer. The surface depicted in Fig.2.3 shows the error curves as functions of the number of cycles for the different cases listed in Table 2.2. The bounds of the error norm during the testing phase are reported in Fig.2.4, confirming what was stated previously. Due to the great discrepancy between the maximum error value of experiment 1 and 2 and the others, only the error bounds of experiments 3, 4, 5, 6 and 7 are given.

2.9.2 Interpolation of analytic functions

Although they may seem to be only a particular class of complex valued functions, analytic functions represent a wide field of interest: all of the complex valued polynomial functions, i.e. transfer functions of linear time-invariant systems, for example, are analytic. Let us consider the following function:

$$f(x + iy) = sin(x)cosh(y) + icos(x)sinh(y) \qquad (2.49)$$

which is analytic in C. In order to approximate this function, different topologies were considered for each of the algorithms, with the aim of selecting the best results as regards both the training error and the generalisation properties. 100 patterns were used to perform the network learning phase and 100 unknown patterns to carry out the testing phase. All the patterns were determined considering the function values in 100 randomly selected complex points $P(x, y)$, with x, $y \in [0, 1]$. The learning parameters were fixed for all the trials to the values:

Algorithm type	A	
Training Number	1	2
Topology	$N_{1,2,1}$	$N_{1,4,1}$
Error	13.96	12.94
No. of parameters	14	26

Algorith type	B	
Training Number	3	4
Topology	$N_{1,2,1}$	$N_{1,4,1}$
Error	0.022	0.029
No. of parameters	14	26

Algorith type	C		
Training Number	5	6	7
Topology	$N_{2,2,2}$	$N_{2,4,2}$	$N_{2,8,2}$
Error	1.5585	0.028	0.0651
No. of parameters	12	22	42

Table 2.2: Function (2.48) approximation

- learning rate = 0.1;

- momentum = 0.01;

- number of cycles = 1100.

The results obtained are given in Table 2.3. As can be observed, algorithm A, which derives from a linear combination of complex analytic functions which is not dense in $C^0(D_n)$, (see theorem 2.6.2), leads in this case to the lowest training error for all the topologies employed. Nevertheless, a density theorem for combinations of functions (2.3) in the set of analytic functions has not yet been proven. In Fig.2.5 the error curves are shown as a function of the experiments carried out; as in the previous examples, the error surface only refers to the last part of the learning phase. In Fig.2.6 the performance of the networks during the testing phase is depicted. As described previously, the error norm on each of the 200 patterns used in the testing phase, for each of the experiments considered, is given. The best performance was obtained in experiment N. 2, corresponding to the network $N_{1,2,1}$ trained with algorithm A. Referring to Section 3.3 it can be deduced that in this case algorithm A easily reaches the best performance, both in the learning and in the testing phase, thus well interpolating the function proposed. This fact should not be a surprise because the A-case uses an activation function belonging to the same class as the function to be interpolated.

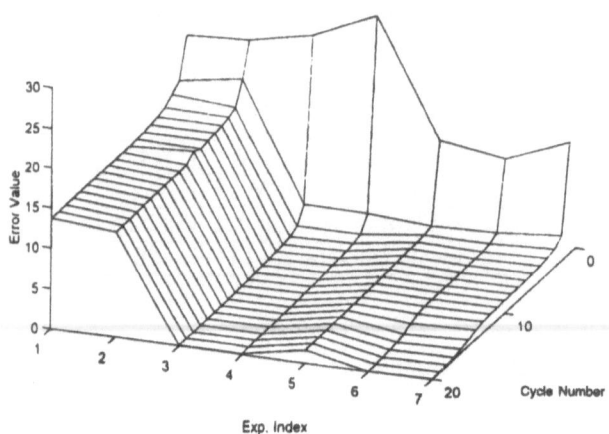

Figure 2.3: *Approximation of function 2.48, learning error trends for the networks of Table 2.2*

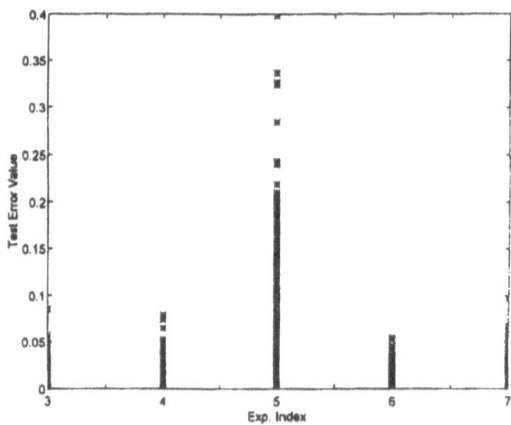

Figure 2.4: *Approximation of function 2.48, testing error absolute values for the networks of Table 2.2*

Algorithm type	A	
Training Number	1	2
Topology	$N_{1,1,1}$	$N_{1,2,1}$
Error	0.0045	0.0036
No. of parameters	8	14

Algorithm type	B		C	
Training Number	3	4	5	6
Topology	$N_{1,1,1}$	$N_{1,2,1}$	$N_{2,2,2}$	$N_{2,4,2}$
Error	0.3482	0.1586	0.2996	0.0457
No. of parameters	8	14	12	22

Table 2.3: Function (2.49) approximation

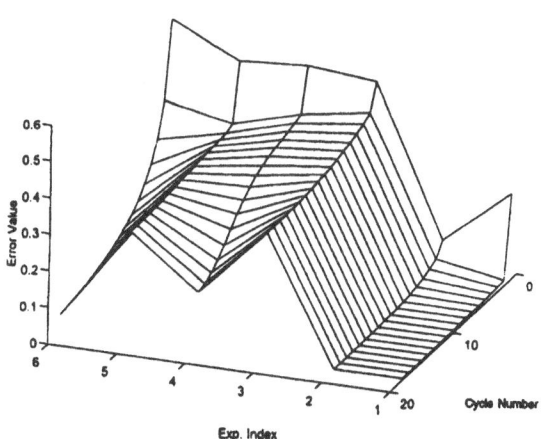

Figure 2.5: *Approximation of 2.49, learning error trends for the networks of Table 2.3*

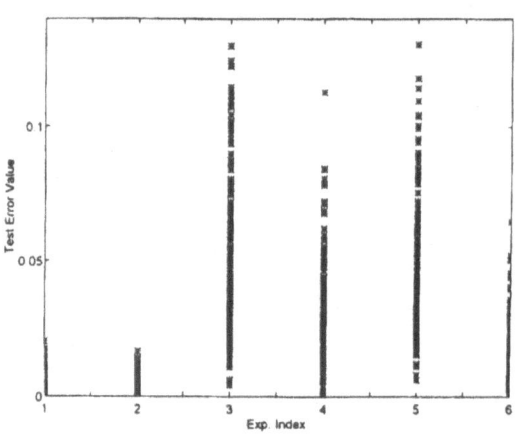

Figure 2.6: *Approximation of function 2.49, testing error absolute values for the networks of Table 2.3*

2.10 Summary

This chapter has introduced the CMLP and analyzed the approximation capabilities of superpositions of complex valued sigmoidal functions on a complex polydisk from a theoretical point of view. Relating the problem to the theory of neural processing, the results obtained led to an analysis of the approximation capabilities of complex feedforward neural networks. More specifically it has been proven that a suitable choice of the network activation functions makes CMLPs universal approximators of complex continuous functions. In particular it is proved that finite linear combinations of type (2.18) can approximate any continuous complex valued function on a general polydisk in C . It is also shown that superpositions of complex sigmoidal functions of type(2.3) can only approximate analytic functions. Whether all of these functions can be approximated by superposition of type (2.3) is a subject for further work. The results derived are useful in order to select the most suitable BP algorithm to solve a given problem with the best performance. The possibility of choosing other activation functions and developing other training algorithms is currently under consideration in order to simplify the network architecture, improve network convergence and obtain better generalization properties.

Further confirmation of the advantages introduced by the CMLP is given when the CMLP is applied to the prediction of some complex valued chaotic time series [6], as will be shown in a following chapter. Moreover, it should be observed that, due to the identity $C = \Re^2$, CMLPs can be used to approach the approximation of real valued multivariable functions, in order to reduce the number of real parameters involved.

From the advantages introduced by the CMLP, the idea of extending the CMLP structure to multi-dimensional domains has arisen. For this purpose, two main strategies can be investigated: the first one generalises the algebra used, while the second one extends the dimensionality of the connections by using real n-dimensional vectors or matrices, The peculiarities of each of the two strategies will be discussed in the following chapters.

Chapter 3

Vectorial neural networks

3.1 Introduction

In most cases neural networks are used to solve problems involving a great number of input and output values. This fact has caused interest in developing neural structures, based on the MLP, built to directly manage vectorial input and output signals. Among these structures two fundamental types have been proposed in literature. This chapter is devoted to describing such vectorial neural networks (VMLPs). The main difference between the two fundamental vectorial neural networks can be found in the structure of the weights connecting the neurons. In both cases, in fact, the neurons accept vectorial inputs and produce vectorial outputs, but the weights connecting the neurons can be vectors, see [26] for the three-dimensional case, or matrices [25]. In particular, in [26] the vector product is considered during the feedforward and back-propagation phases, as will be better explained in the following sections. The next section is devoted to the vectorial MLP with matrix weights.

3.2 VMLP with matrix weights

The VMLP proposed in [25] in 1991 is an MLP with only one hidden layer, in which each neuron is built to manage directly n dimensional vectors. Let us introduce the following notation:

- N :number of input neurons;

- N_1 :number of hidden neurons;

- N_2 :number of output neurons;

- $\epsilon \in \Re$: learning rate;

- $X_i \in \Re^{n*1}$: input vector for the i-th unit $(i = 1, \ldots, N)$;

- $X_{1j} \in \Re^{n*1}$: output vector of the j-th hidden unit $(j = 1, \ldots, N_1)$;

- X_{2k}^d : target vector of the k-th output;

- $X_{2k} \in \Re^{n*1}$: k-th output vector $(k = 1, \ldots, N_2)$;

- $W_{ij} \in \Re^{m*n}$: matrix of the weights connecting the i-th input unit to the j-th hidden unit;

- $W_{1jk} \in \Re^{p*m}$: matrix of the weights connecting the j-th hidden unit to the k-th output unit;

- $\theta_{1j} \in \Re^m$: bias of the j-th hidden unit;

- $\theta_{2k} \in \Re^p$: bias of the k-th output unit;

The activation function for the hidden neurons is built as a vector of real valued sigmoidal functions.
The feedforward phase is defined as follows:

$$X_{1j} = \sigma(O_{1j}) = [\frac{1}{1 + exp(-(o_{1j1} + \theta_{1j1}))}, \ldots, \frac{1}{1 + exp(-(o_{1jm} + \theta_{1jm}))}]^T \quad (3.1)$$

$$X_{2k} = \sigma(O_{1k}) = [\frac{1}{1 + exp(-(o_{1k1} + \theta_{1k1}))}, \ldots, \frac{1}{1 + exp(-(o_{1kp} + \theta_{1kp}))}]^T \quad (3.2)$$

where:

$$O_{1j} = \sum_{i=1}^{N} W_{ij} X_i, \quad j = 1, \ldots, N_1 \quad (3.3)$$

$$O_{2k} = \sum_{i=1}^{N_1} W_{1jk} X_{1j}, \quad k = 1, \ldots, N_2 \quad (3.4)$$

The formulas for weight updating in this structure were again derived following the technique used in the Back Propagation algorithm. The formulas obtained are the following.
For the weights connecting the hidden to the output layer, the following relations hold:

$$\Delta W_{1jk} = \epsilon [\delta_{1k} T_1]^T \quad (3.5)$$

where:

$$\delta_{1k} = [X_{2k}^d - X_{2k}]^T diag[x_{2k1}(1 - x_{2k1}), \ldots, x_{2kp}(1 - x_{2kp})]$$

and T_1 is a tensor $p * m * p$ in which the l-th matrix is:

$$T_{1l} = \begin{bmatrix} 0 \\ \ldots \\ X_{1j}^T \\ 0 \\ \ldots \end{bmatrix}^T \tag{3.6}$$

in which the term X_{1j} is in the l-th row, for $l = 1, \ldots, p$.

The updating formula for the weights connecting the hidden to the input neurons is:

$$\Delta W_{ij} = \epsilon[\delta_j T_2]^T \tag{3.7}$$

where:

$$\delta_j = (\sum_{k=1}^{N_2} \delta_{1k} W_{1jk}) diag[x_{1j1}(1 - x_{1j1}), \ldots, x_{1jm}(1 - x_{1jm})]$$

and T_2 is a $m * n * m$ tensor in which the l-th matrix is:

$$T_{2l} = \begin{bmatrix} 0 \\ \ldots \\ X_i^T \\ 0 \\ \ldots \end{bmatrix}^T \tag{3.8}$$

where the term X_i is in the l-th row, for $l = 1, \ldots, m$.

As regards bias updating the following relations hold:

$$\Delta \theta_{2k} = \epsilon[\delta_{2k}]^T \tag{3.9}$$
$$\Delta \theta_{1j} = \epsilon[\delta_{1j}]^T \tag{3.10}$$

Remark

Taking into account the formulas already shown, it is possible to observe that altough the proposed structure is a compact version of the Multi-layer Perceptron with several input and output signals, it does not introduce any significant advantage from the computational complexity point of view. In fact the number of connections is exactly the same in both cases. More specifically the number of connections needed for the vectorial network (not considering the biases) is:

$$NCv = N * N_1 * m * n + N_1 * N_2 * p * m \tag{3.11}$$

for a vectorial network containing N n-dimensional inputs, N_1 m-dimensional units in the hidden layer, N_2 p-dimensional outputs and $NS = m * N_1 + p * N_2$ sigmoidal functions.

The corresponding real valued MLP therefore needs to contain $N * n$ inputs, $N1 * m$ hidden units and $N_2 * p$ output ones. Thus the connection number will be:

$$NCr = N * n * N_1 * m + N_1 * m * N_2 * p \qquad (3.12)$$

With the same number of sigmoidal functions, we have $NCv = NCr$.

The advantages which derive from the use of this type of vectorial neural network are therefore limited to a mathematical formalism; moreover, several problems remain unsolved, such as determination of the suitable number of hidden neurons as a function of the application at hand, or determination of the correct space dimension for the hidden neurons m.

3.3   3-D MLP in vector algebra

The 3-D vectorial MLP introduced in [26] in 1993 is made-up of weights, biases and input-output signals represented as three-dimensional vectors. Here again the learning algorithm was derived considering a network with one hidden layer. Let us introduce the following notation:

- Ni :number of input neurons;

- Nh :number of hidden neurons;

- No :number of output neurons;

- $W_{ji} = [w_{ji}^x \ w_{ji}^y \ w_{ji}^z]^T \in \Re^3$: weights connecting the i-th input unit to the j-th hidden unit;

- $V_{kj} = [v_{kj}^x \ v_{kj}^y \ v_{kj}^z]^T \in \Re^3$: weight connecting the j-th hidden unit to the k-th output;

- $\Theta_j = [\theta_j^x \ \theta_j^y \ \theta_j^z]^T \in \Re^3$: bias of the j-th hidden neuron;

- $\Gamma_k = [\gamma_k^x \ \gamma_k^y \ \gamma_k^z]^T \in \Re^3$:bias of the k-th output neuron;

- $I_i = [I_i^x \ I_i^y \ I_i^z]^T \in \Re^3$: i-th network input;

- $H_j = [h_j^x \ h_j^y \ h_j^z]^T \in \Re^3$: output of the j-th hidden neuron;

- $O_k = [o_k^x \ o_k^y \ o_k^z]^T \in \Re^3$:k-th network output;

- $T_k = [t_k^x \ t_k^y \ t_k^z]^T \in \Re^3$: target of the k-th output;

- $A_j = [a_j^x \ a_j^y \ a_j^z]^T \in \Re^3$:input to the j-th hidden neuron;

- $B_k = [b_k^x \ b_k^y \ b_k^z]^T \in \Re^3$:input to the k-th output neuron;

The feedforward phase is as follows:

for $j = 1, \ldots, Nh$

$$A_j = \sum_{i=1}^{Ni} (W_{ji} \times I_i) + \Theta_j \tag{3.13}$$

$$H_j = \begin{bmatrix} \dfrac{1}{1 + exp(-a_j^x)} \\ \dfrac{1}{1 + exp(-a_j^y)} \\ \dfrac{1}{1 + exp(-a_j^z)} \end{bmatrix} \tag{3.14}$$

for $k = 1, \ldots, No$

$$B_k = \sum_{i=1}^{Nh} (W_{kj} \times H_j) + \Theta_k \tag{3.15}$$

$$O_k = \begin{bmatrix} \dfrac{1}{1 + exp(-b_k^x)} \\ \dfrac{1}{1 + exp(-b_k^y)} \\ \dfrac{1}{1 + exp(-b_k^z)} \end{bmatrix} \tag{3.16}$$

where 'x' denotes the vector product, given as:

$$A \times B = [a^y b^z - a^z b^y, \ a^z b^x - b^x b^z, \ a^x b^y - a^y b^x]^T$$

It should be observed that the product between two vectors can annihilate even if neither of them is zero; this can degrade network performance, as will be outlined later.

The error between the output and the corresponding target is computed for each pattern as:

$$E = \frac{1}{2} \sum_{k=1}^{No} |(T_k - O_k)|^2 \tag{3.17}$$

The weight updating, computed in order to minimize the error gradient with respect to the weight itself, is given by the following formulas [26]:

$$\Delta V_{kj} = H_j \times \Delta\Gamma_k \tag{3.18}$$

$$\Delta\Gamma_k = \epsilon C_k \Delta_k \tag{3.19}$$

$$\Delta W_{ji} = I_i \times \Delta\Theta_j \tag{3.20}$$

$$\Delta\Theta_j = D_j \sum_{k=1}^{No} (\Delta\Gamma_k \times V_{kj}) \tag{3.21}$$

where:

$$\Delta_k = T_k - O_k \tag{3.22}$$

$$C_k = \begin{bmatrix} (1 - o_k^x)o_k^x & 0 & 0 \\ 0 & (1 - o_k^y)o_k^y & 0 \\ 0 & 0 & (1 - o_k^z)o_k^z \end{bmatrix} \tag{3.23}$$

and:

$$D_j = \begin{bmatrix} (1 - h_j^x)o_h^j & 0 & 0 \\ 0 & (1 - h_j^y)o_j^y & 0 \\ 0 & 0 & (1 - h_j^z)o_h^j \end{bmatrix} \tag{3.24}$$

Remarks.

The structure just introduced allows the computational complexity of the network to be reduced. More specifically, using the same number of sigmoidal interpolating functions, the number of weights is given by:

$$NCv = (Ni * Nh + Nh * No) * 3 \tag{3.25}$$

while the corresponding real MLP with $3 * Ni$ inputs, $3 * Nh$ hidden neurons and $3 * No$ outputs, has a number of weights given by:

$$NCr = 3 * Ni * 3 * Nh + 3 * Nh * 3 * No = 3 * NCv \tag{3.26}$$

It derives that this type of vectorial MLP uses a lower number of parameters to solve the same problem.

Notwithstanding the great computational improvement, this network should be used carefully, due to the properties of vector algebra. In particular, since, as already mentioned, vector algebra is not integer with respect to the vector product, (the product annihilates if the vectors are parallel with each other), some drawbacks can occur.

Let us observe, for example, that some terms in the product $W_{ji} \times I_i$ can annihilate even if the input is not zero and therefore the corresponding input cannot propagate to the next layer. Moreover, during the learning phase, the product $W_{kj} \times H_j$ or the product $H_j \times \Delta\Gamma_k$ may annihilate, even in the presence of a non-zero error, and the corresponding weights therefore remain unchanged. Moreover, from a theoretical point of view, no density theorem has as yet been proven for such a network, so the approximation capabilities of the three-dimensional VMLP have not been investigated. This suggests thinking about a multidimensional neural network developed in a division algebra (i.e. an algebra in which the product of two numbers annihilates only if one of the two factors is zero), in which some density theorems can be proven.

Chapter 4

Quaternion algebra

4.1   Introduction

Quaternion algebra, denoted in the following as H, was invented in 1843 by the Irish mathematician W. R. Hamilton [27], [28] in order to extend the properties of complex numbers to three-dimensional space. In this chapter the fundamental definitions and operations of quaternion algebra are given. A schematic description of the fundamental operations can be found in appendix A. Moreover, a brief overview of some applications of quaternion algebra in different scientific fields is included.

4.2   Fundamental definitions

A quaternion can be considered as a generalized complex number built as the direct sum of a real number q_0 and a three-dimensional vector $\vec{q}$:

$$\mathbf{q} = q_0 + \vec{q} \tag{4.1}$$

In particular q_0 is called a 'scalar quaternion' and $\vec{q}$ is called a 'quaternion vector'. If $(\mathbf{i}, \mathbf{j}, \mathbf{k})$ are the three basis unit vectors representing a three-dimensional orthogonal frame and (q_0, q_1, q_2, q_3) are four real numbers, a quaternion can also be represented as:

$$\mathbf{q} = q_0 + q_1\mathbf{i} + q_2\mathbf{j} + q_3\mathbf{k}. \tag{4.2}$$

In this definition $(\mathbf{i}, \mathbf{j}, \mathbf{k})$ are to be considered as the imaginary unit $i = \sqrt{-1}$ in complex algebra, and the following relations can be stated:

$$\mathbf{i}^2 = \mathbf{j}^2 = \mathbf{k}^2 = -1 \tag{4.3}$$

$$jk = -kj = i \qquad (4.4)$$

$$ki = -ik = j \qquad (4.5)$$

$$ij = -ji = k \qquad (4.6)$$

The following matrix representation of a quaternion is also commonly used:

$$\mathbf{q} = [q_0, \ q_1, \ q_2, \ q_3]^T. \qquad (4.7)$$

The fundamental mathematical operators defined in quaternion algebra are the sum and the product of two quaternions and the inverse operator. They are defined as follows. For any two quaternions:

$$\mathbf{q} = q_0 + q_1 i + q_2 j + q_3 k$$

and:

$$\mathbf{p} = p_0 + p_1 i + p_2 j + p_3 k$$

the sum is defined as follows:

$$\mathbf{q} + \mathbf{p} = (q_0 + p_0) + (q_1 + p_1)i + (q_2 + p_2)j + (q_3 + p_3)k \qquad (4.8)$$

Example
Given:
$q_1 = 1 + 2i + 1j + 2k$
and:
$q_1 = 2 + 3i + 4j + 1k$
we have:
$q_1 + q_2 = 3 + 5i + 5j + 3k$
The quaternion product can be defined as follows:

$$(q_0 + \vec{q}) \otimes (p_0 + \vec{p}) = q_0 p_0 - \vec{p} \cdot \vec{p} + q_0 \vec{p} + p_0 \vec{q} + \vec{q} \times \vec{p} \qquad (4.9)$$

where '.' and 'x' represent the scalar and vector product, respectively, as commonly defined in vector algebra. It has to be pointed out that the product is not commutative and that the product of two 'vector-quaternions' gives a quaternion instead of a vector-quaternion, unlike Vector Algebra.
Example
Given:
$q_1 = 1 + 2i + 1j + 2k$
and:
$q_1 = 2 + 3i + 4j + 1k$
we have:
$q_1 \otimes q_2 = -10 + 0i + 10j + 10k$

and:

$$\mathbf{q_2} \otimes \mathbf{q_1} = -10 + 14\mathbf{i} + 2\mathbf{j} + 0\mathbf{k}$$

A different formalization of the product operator can be given as:

$$\mathbf{q} \otimes \mathbf{p} = \begin{bmatrix} q_0 p_0 - \vec{q}^T \vec{p} \\ \vec{q} p_0 + (q_0 I + \tilde{q}) \vec{p} \end{bmatrix} \tag{4.10}$$

where $\tilde{q}$ denotes the matrix:

$$\tilde{q} = \begin{bmatrix} 0 & -q_3 & q_2 \\ q_3 & 0 & -q_1 \\ -q_2 & q_1 & 0 \end{bmatrix}$$

Even if the the product of two quaternions is not commutative, the associative and distributive properties still hold:

$$\mathbf{q} \otimes (\mathbf{p} \otimes \mathbf{s}) = (\mathbf{q} \otimes \mathbf{p}) \otimes \mathbf{s}$$

$$\mathbf{q} + (\mathbf{p} + \mathbf{s}) = (\mathbf{q} + \mathbf{p}) + \mathbf{s}$$

$$\mathbf{q} \otimes (\mathbf{p} + \mathbf{s}) = \mathbf{q} \otimes \mathbf{p} + \mathbf{q} \otimes \mathbf{s}$$

Owing to the non-commutativity of the product, the right and left-hand product should be differentiated. The conjugate quaternion is derived, as in complex algebra, by changing the sign of the imaginary part, which, in this case, is a vector:

$$q^* = q_o - \vec{q} = q_0 - q_1 \mathbf{i} - q_2 \mathbf{j} - q_3 \mathbf{k} \tag{4.11}$$

The following also holds:

$$\mathbf{q}\mathbf{q}^* = q_0{}^2 + q_1{}^2 + q_2{}^2 + q_3{}^2 \tag{4.12}$$

The modulus is therefore defined, in accordance with the complex algebra, as:

$$|\mathbf{q}| = \sqrt{\mathbf{q}\mathbf{q}^*} \tag{4.13}$$

The inverse of a quaternion is defined as follows:

$$\mathbf{q}^{-1} = \frac{\mathbf{q}^*}{\mathbf{q}^* \otimes \mathbf{q}} \tag{4.14}$$

Example

Given:

$$\mathbf{q_1} = 1 + 2\mathbf{i} + 1\mathbf{j} + 2\mathbf{k}$$

we have:

$\mathbf{q_1}^{-1} = 0.1 - 0.2\mathbf{i} - 0.1\mathbf{j} - 0.2\mathbf{k}$

where:

$\mathbf{q_1}^{-1} \otimes \mathbf{q_1} = \mathbf{q_1} \otimes \mathbf{q_1}^{-1} = 1 + 0\mathbf{i} + 0\mathbf{j} + 0\mathbf{k}$

The definition of the inverse makes quaternion algebra a division algebra. This property allows us to derive an MLP structure in Quaternion Algebra, which is able to avoid the drawbacks encountered in using vector algebra, as discussed previously.

Quaternion algebra is, besides real and complex algebra, the only associative and divisional algebra that can be defined; the following theorems are, in fact, stated [29]:

Theorem 4.2.1 *Each commutative associative and divisional algebra A is isomorphic to $\Re$ or C.*

Theorem 4.2.2 *If A is a real associative and divisional algebra then A is isomorphic to $\Re$, C or H.*

Some Matlab programs which implement the fundamental quaternionic operations are described in Appendix C.

Other useful definitions are:

Definition 4.2.1 *A function $f : H \rightarrow H = f_0(q) + \mathbf{i}f_1(q) + \mathbf{j}f_2(q) + \mathbf{k}f_3(q)$ is continuous if each of its components is a continuous real valued function; it is differentiable if each of its components is differentiable.*

Definition 4.2.2 *Given the operator:*

$$\nabla = \frac{d}{dq_0} + \frac{d}{dq_1}\mathbf{i} + \frac{d}{dq_2}\mathbf{j} + \frac{d}{dq_3}\mathbf{k}$$

the gradient of the function f is defined as $\nabla \otimes f$ or as $f \otimes \nabla$.

Some fundamental definitions on quaternionic differential calcolus can be found in [30], [31].

4.3 Quaternion algebra applications

Quaternion algebra has been used in a wide field of applications, for example in the reformulation of Maxwell equations [30], in the analysis of stresses in three-dimensional objects [32], in the reformulation of kinematic and dynamic equations in three-dimensional space [33], in the trajectory design of manipulators in Cartesian coordinates [34], [35], [36], in molecular dynamics analysis and in digital filter design [37], to name just a few. In all these applications, quaternion algebra has allowed a significant decrease in computational complexity, leading to a linear representation of non-linear phenomena and to a great reduction in the parameters and operations involved. Quaternion algebra allows particular advantages in each of its applications. For example, it has allowed some fundamental systems of partial differential equations, i.e. electromagnetism and fluidodynamics equations to be reformulated in a more compact form, making it possible to derive approximated methods for their solution in suitable boundary conditions [30]. In the field of robotics the analogy between quaternion components and Euler Parameters has been considered in order to represent the joint rotations via quaternion products, thus overcoming some analytical complications typical of mechanical arms analysis. Therefore, the computational complexity of the algorithms of direct and inverse kinematics calculous and trajectory planning can be greatly reduced, also overcoming the redundancy usually introduced through matrix representation. The same reasons led to the introduction of quaternion algebra in molecular chemistry, in order to represent molecular rotations in a simpler computational form.

It is therefore clear that the introduction of an algebra with certain characteristics can allow the efficiency of some analysis and synthesis methods to be increased in various fields.

Chapter 5

MLP in quaternion algebra

5.1 Introduction

As was outlined in the previous chapters, none of the real multi-dimensional MLP-based neural structures guarantees the theoretical results stated for the real MLP, offering in the meanwhile some computational advantages. In particular, for the structure which introduces some computational reduction, no theoretical results have been proven due to the fact that vector algebra is not an integer algebra. This observation led to the idea of developing an MLP-based neural structure in quaternion algebra; the structure is a generalization of the CMLP and, as will be shown below, provides similar performance. In this chapter a new neural network structure based on the Multi-Layer Perceptron and developed in quaternion algebra is therefore introduced. Originally derived to approximate quaternion valued functions, it also shows some advantages in dealing with problems involving a great number of input and output real variables. In the following section some notations used to define the Multi-Layer Perceptron in quaternion algebra, henceforward called HMLP (Hypercomplex Multi-Layer Perceptron) are given. Section 3 describes how the HMLP works. The rules needed to compute the output values for a given input are reported, together with the algorithm for the weight adaptation. In Section 4 the approximation capabilities of the HMLP will be exploited from a theoretical point of view, and some interesting results will be reported which extend the density theorem reported in the previous chapters for the MLP and CMLP.

5.2 Notation

Let us define the HMLP as a neural structure similar to the real MLP, in which all the weights, biases, inputs and outputs are quaternions instead of real numbers. The activation function for the hidden neurons is a suitably chosen quaternion valued function. The following notation is introduced:

- M: number of layers;

- l: layer index;

    (in particular $l = 0$ denotes the input layer and $l = M$ the output one);

- N_l: number of neurons in the l-th layer;

- $\mathbf{X}_n^l = X_{0n}^l + X_{1n}^l \mathbf{i} + X_{2n}^l \mathbf{j} + X_{3n}^l \mathbf{k}$: output of the n-th neuron of the l-th layer;

    (in particular $\mathbf{X}_n^0 = \mathbf{I}_n^0$ for $n = 1, \ldots, N_0$ is the network input and $\mathbf{X}_n^M = \mathbf{Y}_n$ for $n = 1, \ldots, N_M$ is the output)

- $\mathbf{w}_{nm}^l = w_{0nm}^l + w_{1nm}^l \mathbf{i} + w_{2nm}^l \mathbf{j} + w_{3nm}^l \mathbf{k}$: connection weight between the n-th neuron in the l-th layer and the m-th layer in the $l-1$-th layer;

- $\mathbf{S}_n^l = S_{0n}^l + S_{1n}^l \mathbf{i} + S_{2n}^l \mathbf{j} + S_{3n}^l \mathbf{k}$: activation function of the n-th neuron in the l-th layer;

- $\theta_n^l = \theta_{0n}^l + \theta_{1n}^l \mathbf{i} + \theta_{2n}^l \mathbf{j} + \theta_{3n}^l \mathbf{k}$: bias value of the n-th neuron in the l-th layer;

- $\mathbf{t}_n^l = t_{0n}^l + t_{1n}^l \mathbf{i} + t_{2n}^l \mathbf{j} + t_{3n}^l \mathbf{k}$, $n = 1, \ldots, M$: target of the n-th output.

The activation function, chosen in accordance with the CMLP is as follows:

$$\mathbf{X}_n^l = \boldsymbol{\Sigma}(\mathbf{S}_n^l) = \sigma(S_{0n}^l) + \sigma(S_{1n}^l)\mathbf{i} + \sigma(S_{2n}^l)\mathbf{j} + \sigma(S_{3n}^l)\mathbf{k} \qquad (5.1)$$

where:

$$\sigma(x) = \frac{1}{1 + exp(-x)}$$

is the sigmoidal activation function usually adopted in the real MLP. The activation function derivative is:

$$\dot{\boldsymbol{\Sigma}}(\cdot) = \dot{\sigma}(\cdot) + \dot{\sigma}(\cdot)\mathbf{i} + \dot{\sigma}(\cdot)\mathbf{j} + \dot{\sigma}(\cdot)\mathbf{k} \qquad (5.2)$$

where:

$$\dot{\sigma}(\cdot) = \sigma(\cdot)[1 - \sigma(\cdot)]$$

5.3   The learning algorithm for the HMLP

The feedforward phase of the HMLP, which implements transmission of the input signals to the outputs is described as follows:
for $l = 1, \ldots, M$ and $n = 1, \ldots, N_l$

$$\mathbf{S}_n^l = \sum_{m=1}^{N_{l-1}} \mathbf{w}_{nm}^l \otimes \mathbf{X}_m^{l-1} + \theta_n^l \qquad (5.3a)$$

$$\mathbf{X}_n^{l-1} = \boldsymbol{\Sigma}(\mathbf{S}_n^{l-1}) \qquad (5.3b)$$

These relations are used both during the real-time working of the HMLP (once the connection values have been fixed) and during the "learning phase". The learning phase is performed to determine the correct weight values in order to match a given input-output mapping. This means that a set of desired input-output pairs is iteratively presented to the network, and the weights are changed in order to minimize the error between the desired and the actual network output for each input value. Once the input values are presented, the output is computed, as described previously, via (5.3), and it is compared with the target. An error is therefore computed as the square difference between the target and the network output. By repeating the forward phase for the whole set of available patterns, a global square error can be computed as follows:

$$E = \frac{1}{2} \sum_{p=1}^{NPATT} \sum_{n=1}^{N_M} (t_{pn} - \mathbf{X}_{pn}^M)(t_{pn} - \mathbf{X}_{pn}^M)^* \qquad (5.4)$$

where $NPATT$ is the number of learning patterns. The error already computed is used to update the weights in accordance to the gradient descent technique. This means that each weight is changed proportionally to the opposite gradient value of the error with respect to the weight itself. In order to compute the partial derivatives for the weights of the hidden layer, the chain rule should be suitably applied. Such a procedure is classically called the Back Propagation algorithm and it will be extended in the following to quaternion algebra.

In order to give a clear explanation of the mathematical procedure which leads to the hypercomplex back propagation algorithm, a simple HMLP, with one input, one hidden and one output layer is considered.

Let us introduce the following notation:

no :number of output units;

nh :number of units in the hidden layer;

ni :number of input units;

$\mathbf{w}_{nm}^{HO} = w_{0nm}^{HO} + \mathbf{i}w_{1nm}^{HO} + \mathbf{j}w_{2nm}^{HO} + \mathbf{k}w_{3nm}^{HO}$: weight connecting the m-th unit in the hidden layer with the n-th output unit;

$\mathbf{w}_{nm}^{IH} = w_{0nm}^{IH} + \mathbf{i}w_{1nm}^{IH} + \mathbf{j}w_{2nm}^{IH} + \mathbf{k}w_{3nm}^{IH}$: weight connecting the m-th unit in the input layer with the n-th unit in the hidden layer;

$\Theta_n^H = \theta_{0n}^H + \mathbf{i}\theta_{1n}^H + \mathbf{j}\theta_{2n}^H + \mathbf{k}\theta_{3n}^H$: bias of the n-th unit in the hidden layer;

$\Theta_n^O = \theta_{0n}^O + \mathbf{i}\theta_{1n}^O + \mathbf{j}\theta_{2n}^O + \mathbf{k}\theta_{3n}^O$: bias of the n-th unit in the output layer;

Clearly the superscript H denotes the hidden layer, O the output layer and

I the input layer.

The feedforward phase for such a simplified network can be fully described as:

- computation of the hidden unit activation value:

$$\mathbf{S}_n^H = \sum_{m=1}^{ni} \mathbf{w}_{nm}^{IH} \otimes \mathbf{I}_m + \mathbf{\Theta}_n^H = \tag{5.5}$$

$$\sum_{m=1}^{ni} (w_{0nm}^{IH} I_{0m} - w_{0nm}^{IH} I_{0m} - w_{0nm}^{IH} I_{0m} - w_{0nm}^{IH} I_{0m}) + \theta_{0n}^H +$$

$$+\mathbf{i}(\sum_{m=1}^{ni} (w_{1nm}^{IH} I_{0m} + w_{0nm}^{IH} I_{0m} - w_{2nm}^{IH} I_{3m} + w_{3nm}^{IH} I_{2m}) + \theta_{1n}^H) +$$

$$+\mathbf{j}(\sum_{m=1}^{ni} (w_{2nm}^{IH} I_{0m} + w_{0nm}^{IH} I_{2m} + w_{1nm}^{IH} I_{3m} - w_{3nm}^{IH} I_{1m}) + \theta_{2n}^H) +$$

$$+\mathbf{k}(\sum_{m=1}^{ni} (w_{3nm}^{IH} I_{0m} + w_{0nm}^{IH} I_{3m} - w_{1nm}^{IH} I_{2m} + w_{2nm}^{IH} I_{1m}) + \theta_{3n}^H)$$

- computation of the hidden unit output:

$$\mathbf{X}_n^H = \mathbf{\Sigma}(\mathbf{S}_n^H) = \sigma(S_{0n}^H) + \mathbf{i}\sigma(S_{1n}^H) + \mathbf{j}\sigma(S_{2n}^H) + \mathbf{k}\sigma(S_{3n}^H) \tag{5.6}$$

- computation of the output unit activation value:

$$\mathbf{S}_n^O = \sum_{m=1}^{nh} \mathbf{w}_{nm}^{HO} \otimes \mathbf{X}_m^H + \mathbf{\Theta}_n^O = \tag{5.7}$$

$$\sum_{m=1}^{nh} (w_{0nm}^{HO} X_{0m}^H - w_{0nm}^{HO} X_{0m}^H - w_{0nm}^{HO} X_{0m}^H - w_{0nm}^{HO} X_{0m}^H) + \theta_{0n}^O +$$

$$\mathbf{i}(\sum_{m=1}^{nh} (w_{1nm}^{HO} X_{0m}^H + w_{0nm}^{HO} X_{0m}^H - w_{2nm}^{HO} X_{3m}^H + w_{3nm}^{HO} X_{2m}^H) + \theta_{1n}^O) +$$

$$\mathbf{j}(\sum_{m=1}^{nh} (w_{2nm}^{HO} X_{0m}^H + w_{0nm}^{HO} X_{2m}^H + w_{1nm}^{HO} X_{3m}^H - w_{3nm}^{HO} X_{1m}^H) + \theta_{2n}^O) +$$

$$\mathbf{k}(\sum_{m=1}^{nh} (w_{3nm}^{HO} X_{0m}^H + w_{0nm}^{HO} X_{3m}^H - w_{1nm}^{HO} X_{2m}^H + w_{2nm}^{HO} X_{1m}^H) + \theta_{3n}^O)$$

- computation of the *n*-th output:

$$\mathbf{X}_n^O = \mathbf{Y}_n = \mathbf{\Sigma}(\mathbf{S}_n^O) = \sigma(S_{0n}^O) + \mathbf{i}\sigma(S_{1n}^O) + \mathbf{j}\sigma(S_{2n}^O) + \mathbf{k}\sigma(S_{3n}^O) \tag{5.8}$$

The error function which should be minimized is therefore:

$$E = \frac{1}{2} \sum_{n=1}^{no} [(t_{0n} - Y_{0n})^2 + (t_{1n} - Y_{1n})^2 + (t_{2n} - Y_{2n})^2 + (t_{3n} - Y_{3n})^2] \quad (5.9)$$

In order to derive the weight updating formulas it is necessary to compute the error function gradient with respect to each weight connecting the hidden and the output layer and subsequently the gradient with respect to the weight connecting the hidden and the input layer. The same procedure will be applied for the bias update. Since the error function is a real valued function we have:

$$\nabla E_{\mathbf{w}_{nm}^{HO}} = \frac{\partial E}{\partial w_{0nm}^{HO}} + \mathbf{i}\frac{\partial E}{\partial w_{1nm}^{HO}} + \mathbf{j}\frac{\partial E}{\partial w_{2nm}^{HO}} + \mathbf{k}\frac{\partial E}{\partial w_{3nm}^{HO}} \quad (5.10)$$

Each term is computed applying the chain rule. For the first term of 5.10 we get:

$$\frac{\partial E}{\partial w_{0nm}^{HO}} = \frac{\partial E}{\partial S_{0n}^{O}} \frac{\partial S_{0n}^{O}}{\partial w_{0nm}^{HO}} + \frac{\partial E}{\partial S_{1n}^{O}} \frac{\partial S_{1n}^{O}}{\partial w_{0nm}^{HO}} + \frac{\partial E}{\partial S_{2n}^{O}} \frac{\partial S_{2n}^{O}}{\partial w_{0nm}^{HO}} + \frac{\partial E}{\partial S_{3n}^{O}} \frac{\partial S_{3n}^{O}}{\partial w_{0nm}^{HO}} \quad (5.11)$$

Taking 5.7, 5.8 and 5.9 into account, the following relations are obtained:

$$\frac{\partial E}{\partial S_{0n}^{O}} = \frac{\partial E}{\partial Y_{0n}} \frac{\partial Y_{0n}}{\partial S_{0n}^{O}} = -(t_{0n} - Y_{0n})\dot{\sigma}(S_{0n}^{O}) \quad (5.12)$$

$$\frac{\partial E}{\partial S_{1n}^{O}} = \frac{\partial E}{\partial Y_{1n}} \frac{\partial Y_{1n}}{\partial S_{1n}^{O}} = -(t_{1n} - Y_{1n})\dot{\sigma}(S_{1n}^{O}) \quad (5.13)$$

$$\frac{\partial E}{\partial S_{2n}^{O}} = \frac{\partial E}{\partial Y_{2n}} \frac{\partial Y_{2n}}{\partial S_{2n}^{O}} = -(t_{2n} - Y_{2n})\dot{\sigma}(S_{2n}^{O}) \quad (5.14)$$

$$\frac{\partial E}{\partial S_{3n}^{O}} = \frac{\partial E}{\partial Y_{3n}} \frac{\partial Y_{3n}}{\partial S_{3n}^{O}} = -(t_{3n} - Y_{3n})\dot{\sigma}(S_{3n}^{O}) \quad (5.15)$$

$$\frac{\partial S_{0n}^{O}}{\partial w_{0nm}^{HO}} = X_{0m}^{H} \quad (5.16)$$

$$\frac{\partial S_{1n}^{O}}{\partial w_{0nm}^{HO}} = X_{1m}^{H} \quad (5.17)$$

$$\frac{\partial S_{2n}^{O}}{\partial w_{0nm}^{HO}} = X_{2m}^{H} \quad (5.18)$$

$$\frac{\partial S_{3n}^{O}}{\partial w_{0nm}^{HO}} = X_{3m}^{H} \quad (5.19)$$

From these formulas we get:

$$\frac{\partial E}{\partial w_{0nm}^{HO}} = -[(t_{0n} - Y_{0n})\dot{\sigma}(S_{0n}^O)X_{0m}^H + (t_{1n} - Y_{1n})\dot{\sigma}(S_{1n}^O)X_{1m}^H + \quad (5.20)$$

$$(t_{2n} - Y_{2n})\dot{\sigma}(S_{2n}^O)X_{2m}^H + (t_{3n} - Y_{3n})\dot{\sigma}(S_{3n}^O)X_{3m}^H]$$

For second term of the 5.10 we get:

$$\frac{\partial E}{\partial w_{1nm}^{HO}} = \frac{\partial E}{\partial S_{0n}^O}\frac{\partial S_{0n}^O}{\partial w_{1nm}^{HO}} + \frac{\partial E}{\partial S_{1n}^O}\frac{\partial S_{1n}^O}{\partial w_{1nm}^{HO}} + \frac{\partial E}{\partial S_{2n}^O}\frac{\partial S_{2n}^O}{\partial w_{1nm}^{HO}} + \frac{\partial E}{\partial S_{3n}^O}\frac{\partial S_{3n}^O}{\partial w_{1nm}^{HO}}] \quad (5.21)$$

and then, applying the chain rule once more and considering 5.7, 5.8 and 5.9:

$$\frac{\partial E}{\partial w_{1nm}^{HO}} = -[(t_{0n} - Y_{0n})\dot{\sigma}(S_{0n}^O)(-X_{1m}^H) + (t_{1n} - Y_{1n})\dot{\sigma}(S_{1n}^O)X_{0m}^H + \quad (5.22)$$

$$(t_{2n} - Y_{2n})\dot{\sigma}(S_{2n}^O)X_{3m}^H + (t_{3n} - Y_{3n})\dot{\sigma}(S_{3n}^O)(-X_{2m}^H)$$

in fact:

$$\frac{\partial S_{0n}^O}{\partial w_{1nm}^{HO}} = -X_{1m}^H \quad (5.23)$$

$$\frac{\partial S_{1n}^O}{\partial w_{1nm}^{HO}} = X_{0m}^H \quad (5.24)$$

$$\frac{\partial S_{2n}^O}{\partial w_{1nm}^{HO}} = X_{3m}^H \quad (5.25)$$

$$\frac{\partial S_{3n}^O}{\partial w_{1nm}^{HO}} = -X_{2m}^H \quad (5.26)$$

In the same way for the third term of 5.10 we obtain:

$$\frac{\partial E}{\partial w_{2nm}^{HO}} = \frac{\partial E}{\partial S_{0n}^O}\frac{\partial S_{0n}^O}{\partial w_{2nm}^{HO}} + \frac{\partial E}{\partial S_{1n}^O}\frac{\partial S_{1n}^O}{\partial w_{2nm}^{HO}} + \frac{\partial E}{\partial S_{2n}^O}\frac{\partial S_{2n}^O}{\partial w_{2nm}^{HO}} + \frac{\partial E}{\partial S_{3n}^O}\frac{\partial S_{3n}^O}{\partial w_{2nm}^{HO}}] \quad (5.27)$$

from which:

$$\frac{\partial E}{\partial w_{2nm}^{HO}} = -[(t_{0n} - Y_{0n})\dot{\sigma}(S_{0n}^O)(-X_{2m}^H) + (t_{1n} - Y_{1n})\dot{\sigma}(S_{1n}^O)(-X_{3m}^H) + \quad (5.28)$$

$$(t_{2n} - Y_{2n})\dot{\sigma}(S_{2n}^O)X_{0m}^H + (t_{3n} - Y_{3n})\dot{\sigma}(S_{3n}^O)S_{1m}^H$$

Finally, for the last term of 5.10 the following relation holds:

$$\frac{\partial E}{\partial w_{3nm}^{HO}} = \frac{\partial E}{\partial S_{0n}^O}\frac{\partial S_{0n}^O}{\partial w_{3nm}^{HO}} + \frac{\partial E}{\partial S_{1n}^O}\frac{\partial S_{1n}^O}{\partial w_{3nm}^{HO}} + \frac{\partial E}{\partial S_{2n}^O}\frac{\partial S_{2n}^O}{\partial w_{3nm}^{HO}} + \frac{\partial E}{\partial S_{3n}^O}\frac{\partial S_{3n}^O}{\partial w_{3nm}^{HO}}] \quad (5.29)$$

from which:

$$\frac{\partial E}{\partial w_{2nm}^{HO}} = -[(t_{0n} - Y_{0n})\dot{\sigma}(S_{0n}^O)(-X_{3m}^H) + (t_{1n} - Y_{1n})\dot{\sigma}(S_{1n}^O)X_{2m}^H + \quad (5.30)$$

$$(t_{2n} - Y_{2n})\dot{\sigma}(S_{2n}^O)(-X_{1m}^H + (t_{3n} - Y_{3n})\dot{\sigma}(S_{3n}^O)X_{0m}^H$$

The formula obtained, which allows us to update the weights connecting the hidden units with the neurons of the output layer is therefore:

$$\Delta w_{nm}^{HO} = \epsilon(-\nabla E_{\mathbf{w}_{nm}^{HO}}) = \epsilon[(t_n - \mathbf{Y}_n) \odot \mathbf{\Sigma}(\mathbf{S}_n^O)] \otimes \mathbf{X}_m^{H*} = \epsilon \delta_n^O \otimes \mathbf{X}_m^{H*} \quad (5.31)$$

where $\odot$ denotes the component-by-component product of two quaternions, and $\otimes$ denotes the quaternionic product while $\epsilon \in \Re^+$ is the learning rate.
In order to derive the updating formulas for the weights connecting the hidden layer units with the input, the following gradient should be computed:

$$\nabla E_{\mathbf{w}_{nm}^{IH}} = \frac{\partial E}{\partial w_{0nm}^{IH}} + \mathbf{i}\frac{\partial E}{\partial w_{1nm}^{IH}} + \mathbf{j}\frac{\partial E}{\partial w_{2nm}^{IH}} + \mathbf{k}\frac{\partial E}{\partial w_{3nm}^{IH}} \quad (5.32)$$

Here again, each term is obtained by applying the chain rule:
Therefore we get:

$$\frac{\partial E}{\partial w_{0nm}^{IH}} = \sum_{k=1}^{no}[\frac{\partial E}{\partial S_{0k}^O}\frac{\partial S_{0k}^O}{\partial w_{0nm}^{IH}} + \frac{\partial E}{\partial S_{1k}^O}\frac{\partial S_{1k}^O}{\partial w_{0nm}^{IH}} + \frac{\partial E}{\partial S_{2k}^O}\frac{\partial S_{2k}^O}{\partial w_{0nm}^{IH}} + \frac{\partial E}{\partial S_{3k}^O}\frac{\partial S_{3k}^O}{\partial w_{0nm}^{IH}}]$$

$$(5.33)$$

where:

$$\frac{\partial S_{0k}^O}{\partial w_{0nm}^{IH}} = \frac{\partial S_{0k}^O}{\partial X_{0n}^H}\frac{\partial X_{0n}^H}{\partial w_{0nm}^{IH}} + \frac{\partial S_{0k}^O}{\partial X_{1n}^H}\frac{\partial X_{1n}^H}{\partial w_{0nm}^{IH}} + \frac{\partial S_{0k}^O}{\partial X_{2n}^H}\frac{\partial X_{2n}^H}{\partial w_{0nm}^{IH}} + \frac{\partial S_{0k}^O}{\partial X_{3n}^H}\frac{\partial X_{3n}^H}{\partial w_{0nm}^{IH}} \quad (5.34)$$

$$= w_{0kn}^{HO}\dot{\sigma}(S_{0n}^H)I_{0m} + (-w_{1kn}^{HO})\dot{\sigma}(S_{1n}^H)I_{1m} + (-w_{2kn}^{HO})\dot{\sigma}(S_{2n}^H)I_{2m} + (-w_{3kn}^{HO})\dot{\sigma}(S_{3n}^H)I_{3m}$$

In the same way:

$$\frac{\partial S_{1k}^O}{\partial w_{0nm}^{IH}} = \quad (5.35)$$

$$w_{1km}^{HO}\dot{\sigma}(S_{0n}^H)I_{0m} + (w_{0km}^{HO})\dot{\sigma}(S_{1n}^H)I_{1m} + (-w_{3km}^{HO})\dot{\sigma}(S_{2m}^H)I_{2m} + (w_{2km}^{HO})\dot{\sigma}(S_{3n}^H)I_{3m}$$

$$\frac{\partial S_{2k}^O}{\partial w_{0nm}^{IH}} = \quad (5.36)$$

$$w_{2km}^{HO}\dot{\sigma}(S_{0n}^H)I_{0m} + (w_{3km}^{HO})\dot{\sigma}(S_{1n}^H)I_{1m} + (w_{0km}^{HO})\dot{\sigma}(S_{2n}^H)I_{2m} + (-w_{1km}^{HO})\dot{\sigma}(S_{3n}^H)I_{3m}$$

$$\frac{\partial S_{3k}^O}{\partial w_{0nm}^{IH}} = \quad (5.37)$$

$$w_{3km}^{HO}\dot{\sigma}(S_{0n}^H)I_{0m} + (-w_{2km}^{HO})\dot{\sigma}(S_{1n}^H)I_{1m} + (w_{1km}^{HO})\dot{\sigma}(S_{2n}^H)I_{2m} + (w_{0km}^{HO})\dot{\sigma}(S_{3n}^H)I_{3m}$$

The following hold:

$$\frac{\partial E}{\partial S_{0k}^O} = -(t_{0k} - Y_{0k})\dot{\sigma}(S_{0k}^O) \qquad (5.38)$$

$$\frac{\partial E}{\partial S_{1k}^O} = -(t_{1k} - Y_{1k})\dot{\sigma}(S_{1k}^O) \qquad (5.39)$$

$$\frac{\partial E}{\partial S_{2k}^O} = -(t_{2k} - Y_{2k})\dot{\sigma}(S_{2k}^O) \qquad (5.40)$$

$$\frac{\partial E}{\partial S_{3k}^O} = -(t_{3k} - Y_{3k})\dot{\sigma}(S_{3k}^O) \qquad (5.41)$$

Following the previous steps the other terms of 5.32 can be computed as follows:

$$\frac{\partial E}{\partial w_{1nm}^{IH}} = \sum_{k=1}^{no} \left[\frac{\partial E}{\partial S_{0k}^O} \frac{\partial S_{0k}^O}{\partial w_{1nm}^{IH}} + \frac{\partial E}{\partial S_{1k}^O} \frac{\partial S_{1k}^O}{\partial w_{1nm}^{IH}} + \frac{\partial E}{\partial S_{2k}^O} \frac{\partial S_{2k}^O}{\partial w_{1nm}^{IH}} + \frac{\partial E}{\partial S_{3k}^O} \frac{\partial S_{3k}^O}{\partial w_{1nm}^{IH}} \right]$$
$$(5.42)$$

$$\frac{\partial E}{\partial w_{2nm}^{IH}} = \sum_{k=1}^{no} \left[\frac{\partial E}{\partial S_{0k}^O} \frac{\partial S_{0k}^O}{\partial w_{2nm}^{IH}} + \frac{\partial E}{\partial S_{1k}^O} \frac{\partial S_{1k}^O}{\partial w_{2nm}^{IH}} + \frac{\partial E}{\partial S_{2k}^O} \frac{\partial S_{2k}^O}{\partial w_{2nm}^{IH}} + \frac{\partial E}{\partial S_{3k}^O} \frac{\partial S_{3k}^O}{\partial w_{2nm}^{IH}} \right]$$
$$(5.43)$$

$$\frac{\partial E}{\partial w_{3nm}^{IH}} = \sum_{k=1}^{no} \left[\frac{\partial E}{\partial S_{0k}^O} \frac{\partial S_{0k}^O}{\partial w_{3nm}^{IH}} + \frac{\partial E}{\partial S_{1k}^O} \frac{\partial S_{1k}^O}{\partial w_{3nm}^{IH}} + \frac{\partial E}{\partial S_{2k}^O} \frac{\partial S_{2k}^O}{\partial w_{3nm}^{IH}} + \frac{\partial E}{\partial S_{3k}^O} \frac{\partial S_{3k}^O}{\partial w_{3nm}^{IH}} \right]$$
$$(5.44)$$

Taking the previous terms into account, the final updating formula is obtained as:

$$\Delta \mathbf{w}_{nm}^{IH} = \epsilon(-\nabla E_{\mathbf{w}_{nm}^{IH}}) = \epsilon \delta_n^H \otimes \mathbf{I}_m^* \qquad (5.45)$$

where

$$\delta_n^H = \left[\sum_{k=1}^{no} \mathbf{W}_{kn}^{*HO} \otimes \delta_k^O \right] \cdot \dot{\sigma}(\mathbf{S}_n^H) \qquad (5.46)$$

Remembering that the biases correspond to weights from units with a constant output equal to one, we get:

$$\Delta \theta_n^O = \epsilon \delta_n^O \qquad (5.47)$$

$$\Delta \theta_n^H = \epsilon \delta_n^H \qquad (5.48)$$

The same approach is adopted to determine the adaptation rules for an HMLP with more than one hidden layer.

The formulas obtained are schematically reported below, with the notation introduced at the beginning of the chapter.

Learning algorithm for the HMLP
for $l = 1, \ldots, M$ and $n = 1, \ldots, N_l$

$$\begin{cases} e_n^l = t_n - \mathbf{X}_n^M & \text{if } l = M \\ e_n^l = \sum_{h=1}^{N_{l+1}} \mathbf{w}_{hn}^{*l+1} \otimes \delta_h^{l+1} & \text{if } l = M-1, \ldots, 1 \end{cases} \qquad (5.49)$$

where:

$$\delta_h^{l+1} = e_{0n}^{l+1} \dot{\sigma}(S_{0n}^{l+1}) + e_{1n}^{l+1} \dot{\sigma}(S_{1n}^{l+1})\mathbf{i} + e_{2n}^{l+1} \dot{\sigma}(S_{2n}^{l+1})\mathbf{j} + e_{3n}^{l+1} \dot{\sigma}(S_{3n})^{l+1}\mathbf{k} \quad (5.50)$$

$$\mathbf{w}_{nm}^l(k+1) = \mathbf{w}_{nm}^l(k) + \epsilon \delta_n^l \otimes \mathbf{X}_m^{*(l-1)} \qquad (5.51)$$
$$\theta_n^l(k+1) = \theta_n^l(k) + \epsilon \delta_n^l \qquad (5.52)$$

where the index k denotes the k-th iteration of the learning algorithm, ϵ the learning rate and $m = 0, \ldots, N_{l-1}$.

Remark:
It should be observed that this algorithm reduces to the real B.P. algorithm and to the learning algorithm for the CMLP with a not analytic activation function (2.21) when real and complex signals are involved.

5.4 Multi-layer perceptrons in real or in quaternion algebra: a comparison

In this section a comparison is made between the HMLP and the real MLP as regards the number of parameters employed.
Let us consider an HMLP with only one hidden layer, Ni inputs, Nh hidden neurons and No outputs. Let us denote with NCc the number of real weights, i.e. the number of weights obtained considering that each quaternionic weight corresponds to four real weights. This number is given by:

$$NCc = (Ni * Nh + Nh * No) * 4 \qquad (5.53)$$

The equivalent network, considering the same number of interpolating functions, comprises of $4 * Ni$ inputs, $4 * Nh$ hidden neurons and $4 * No$ output units. The number of weights will therefore be:

$$NCr = 4 * Ni * 4 * Nh + 4 * Nh * 4 * No = 4 * NCc \qquad (5.54)$$

Considering the same number of real sigmoidal functions embedded in the network, the HMLP needs a quarter of the number of parameters required by the real MLP (the number of biases being equal in both structures). Although it has been observed that the number of interpolating functions needed for the HMLP to reach the same performance as the real MLP is slightly greater, the real parameter saving remains considerable.

5.5   Density theorem for the HMLP

Use of the quaternionic structures mentioned in the previous chapters should be justified from a theoretical point of view in order to assure the existence of a solution, i.e. a set of connection values able to accomplish the desired task. Such an analysis has been performed for the real MLP by several authors, [15], [19], [20], and was subsequently extended to CMLPs [22], [23], [44], as reported in the previous chapters. In this section a density theorem for continuous functions $f : X \rightarrow H$, where X is a compact subset of H^n, will be proven. As for the real MLP [15] and the CMLP [22] this result states that HMLPs with the activation function (5.1) are universal interpolators of continuous quaternion valued functions.

In order to prove the theorem the following notation is introduced:

Let $\sigma : \Re \rightarrow \Re$ be the real valued sigmoidal function:

$$\sigma(t) = \frac{1}{1 + exp(-t)}$$

and let $\mathbf{q} = q_0 + \mathbf{i}q_1 + \mathbf{j}q_2 + \mathbf{k}q_3 \in H$.

Let us consider the function $f : X \rightarrow H$:

$$f(\mathbf{q}) = \sigma(q_0) + \mathbf{i}\sigma(q_1) + \mathbf{j}\sigma(q_2) + \mathbf{k}\sigma(q_3) \tag{5.55}$$

and the following functions from H to $\Re$:

$$R(\mathbf{q}) = q_0 \quad I(\mathbf{q}) = q_1 \quad J(\mathbf{q}) = q_2 \quad K(\mathbf{q}) = q_3$$

It follows that:

$$f = \sigma \circ R + \mathbf{i}\sigma \circ I + \mathbf{j}\sigma \circ J + \mathbf{k}\sigma \circ K$$

where o represents the function composition operator.

Let H^n be the set of quaternion n-uples:

$$H^n = \{\bar{\mathbf{x}} = (\mathbf{x}_1, \ldots, \mathbf{x}_n)^T \quad \mathbf{x}_i \in H\}$$

Having fixed $\bar{\mathbf{y}} \in H^n$, the following function can be considered:

$$\bar{\mathbf{y}}^T \cdot \bar{\mathbf{x}} = \mathbf{y}_1 \otimes \mathbf{x}_1 + \ldots + \mathbf{y}_n \otimes \mathbf{x}_n$$

where the operator '·' represents the scalar product between vectors. The function just defined is $\Re$ -linear, i.e. :

$$\bar{\mathbf{y}}^T \cdot (\bar{\mathbf{x}} + \bar{\mathbf{x}}') = \bar{\mathbf{y}}^T \cdot \bar{\mathbf{x}} + \bar{\mathbf{y}}^T \cdot \bar{\mathbf{x}}'$$

$$\bar{\mathbf{y}}^T \cdot (\lambda \bar{\mathbf{x}}) = \lambda \bar{\mathbf{y}}^T \cdot \bar{\mathbf{x}} \quad \forall \lambda \in \Re$$

It is now possible to prove the following density theorem:

Theorem 5.5.1 *Let $X \subset H^n$ be a compact subset and let $g : X \to H$ be a continuous function. Then $\forall \epsilon > 0$ there exist some coefficients:*

$$\alpha_1, \ldots \alpha_n \in \Re$$

some vectors:

$$\bar{\mathbf{y}}_1, \ldots, \bar{\mathbf{y}}_N \in H^n$$

and some quaternions:

$$\Theta_1, \ldots, \Theta_N \in H$$

such that:

$$sup_{\bar{\mathbf{x}} \in X} \| \bar{g}(\bar{\mathbf{x}}) - \sum_{i=1}^{N} \alpha_i f(\bar{\mathbf{y}}_i^T \cdot \bar{\mathbf{x}} + \bar{\Theta}_i) \| < \epsilon. \tag{5.56}$$

In other words, the real vector space:

$$S = \{\sum_{i=1}^{N} \alpha_i f(\bar{\mathbf{y}}_i^T \cdot \bar{\mathbf{x}} + \Theta_i), \text{ where } N \text{ is a natural number}, \alpha_i \in \Re, \bar{\mathbf{y}}_i \in H^n, \Theta_i \in H\}$$

is dense in $C^0(X, H)$, the space of continuous functions $X \to H$ with the norm

$$\| g \| = sup_{\bar{\mathbf{x}} \in X} |g(\bar{\mathbf{x}})|$$

Proof:
In order to prove this theorem the following lemma is introduced:

Lemma 5.5.1 *Let $\phi : H^n \to \Re$ be a linear function.*
Then $\bar{\mathbf{y}}_R, \bar{\mathbf{y}}_I, \bar{\mathbf{y}}_J, \bar{\mathbf{y}}_K \in H^n$ are univocally determined such that, $\forall \bar{\mathbf{x}} \in H^n$, it follows that:

$$\phi(\bar{\mathbf{x}}) = R(\bar{\mathbf{y}}_R^T \cdot \bar{\mathbf{x}}) = I(\bar{\mathbf{y}}_I^T \cdot \bar{\mathbf{x}}) = J(\bar{\mathbf{y}}_J^T \cdot \bar{\mathbf{x}}) = K(\bar{\mathbf{y}}_K^T \cdot \bar{\mathbf{x}})$$

Proof of Lemma 5.7.1
The proof is only given for the representation $\phi(\bar{\mathbf{x}}) = R(\bar{\mathbf{y}}_R^T \cdot \bar{\mathbf{x}})$, as it is easy to extend it to all other cases.
Let us define the following application:

$$\Phi : H^n \to \{\phi : H^n \to \Re : \phi \text{ is linear}\}$$

defined as: $\Phi(\mathbf{y}) = \phi$ such that $\phi(\bar{\mathbf{x}}) = R(\bar{\mathbf{y}}^T \cdot \bar{\mathbf{x}}) \forall \bar{\mathbf{x}} \in H^n$.
It has to be proven that Φ is a one-to-one correspondence.

It should be observed that Φ is a linear application between vector spaces of the same dimension. To prove the lemma it is therefore sufficient to show that Φ is injective.

This means that if $\bar{\mathbf{y}} \in H^n$ is such that $R(\bar{\mathbf{y}}_R^T \cdot \bar{\mathbf{x}}) = 0 \quad \forall \bar{\mathbf{x}} \in H^n$, it follows that $\bar{\mathbf{y}} = 0$.

Let us prove what has been stated.

Let us consider $\bar{\mathbf{y}} = (\mathbf{y}_1, \ldots, \mathbf{y}_n)^T$ and take into account, for the sake of simplicity, the first component $\mathbf{y}_1 \in H$.

Owing to the fact that:

$$0 = R(\bar{\mathbf{y}}^T \cdot \mathbf{e}_1) = R(\bar{\mathbf{y}}^T \cdot (i\mathbf{e}_1)) = R(\bar{\mathbf{y}}^T \cdot (j\mathbf{e}_1)) = R(\bar{\mathbf{y}}^T \cdot (k\mathbf{e}_1))$$

if $\mathbf{e}_1 = (1 \, 0 \, \ldots \, 0)^T$ it follows that:

$$R(\mathbf{y}_1) = I(\mathbf{y}_1) = J(\mathbf{y}_1) = K(\mathbf{y}_1) = 0 \quad \rightarrow \quad \mathbf{y}_1 = 0$$

In the same way it can be shown that $\mathbf{y}_i = 0$ when $i = 2, \ldots, n$ and therefore $\bar{\mathbf{y}} = 0$

The lemma has therefore been proven.

Proof of Theorem 5.7.1

Let us consider the following continuous function $g : X \rightarrow H$ together with its four components:

$$g = g_0 + ig_1 + jg_2 + kg_3 = R \circ g + iI \circ g + jJ \circ g + kK \circ g$$

If S can separately approximate the functions:

$$R \circ g, \quad iI \circ g, \quad jJ \circ g, \quad kK \circ g$$

then of course S can approximate g.

Let us first consider the real part of g, denoted as $R \circ g : X \rightarrow \Re$.

From Cybenko's density theorem in the real field applied to $C^0(X, \Re)$, (where X can be seen as a compact subset of $\Re^{4n}$) it follows that there exist some real coefficients $a_1, \ldots, a_N$, some sigmoidal real valued functions $\phi_1, \ldots \phi_N$ on $\Re^{4n} = H^n$ and some real parameters $\theta_1, \ldots, \theta_N$ such that:

$$\left| R(g(\bar{\mathbf{x}})) - \sum_{i=1}^{N} a_i \sigma(\phi_i(\bar{\mathbf{x}}) + \theta_i) \right| < \epsilon$$

$forall \bar{\mathbf{x}} \in X$.

For the lemma already proven we have $\phi_i(\bar{\mathbf{x}}) = R(\bar{\mathbf{y}}_i^T \cdot \bar{\mathbf{x}})$ (where $\bar{\mathbf{x}}$ is now a vector of H^n and $\bar{\mathbf{y}}_i \in H^n$).

Let us consider:

$$\Theta_i(\lambda) = \theta_i + i\lambda + j\lambda + k\lambda$$

with $\lambda \in \Re$.

The function f can therefore be represented as:

$$f(\bar{\mathbf{y}}_i^T \cdot \bar{\mathbf{x}} + \Theta_i(\lambda)) = \sigma(\phi_i(\bar{\mathbf{x}}) + \Theta_i) + \mathbf{i}I(\bar{\mathbf{y}}_i^T \cdot \bar{\mathbf{x}} + \lambda) + \mathbf{j}J(\bar{\mathbf{y}}_i^T \cdot \bar{\mathbf{x}} + \lambda) + \mathbf{k}K(\bar{\mathbf{y}}_i^T \cdot \bar{\mathbf{x}} + \lambda)$$

Since $\bar{\mathbf{x}}$ is in the compact set X and $\bar{\mathbf{y}}_i$ are fixed, the imaginary parts of the product $\bar{\mathbf{y}}_i^T \cdot \bar{\mathbf{x}}$ are bounded.

Then, if $\lambda \to -\infty$ the imaginary parts of the function $f(\bar{\mathbf{y}}_i^T \cdot \bar{\mathbf{x}} + \Theta_i(\lambda))$ converge uniformly to 0 and the following convergence:

$$lim_{\lambda \to -\infty} f(\bar{\mathbf{y}}_i^T \cdot \bar{\mathbf{x}} + \Theta_i(\lambda)) = \sigma(\phi_i(\bar{\mathbf{x}}) + \Theta_i)$$

is uniform, for all $i = 1, \ldots, N$.

It follows that a value of λ exists such that:

$$|\sum_{i=1}^{N} \alpha_i \sigma(\phi_i(\bar{\mathbf{x}}) + \theta_i) - \sum_{i=1}^{N} \alpha_i f_i(\bar{\mathbf{y}}_i^T \bar{\mathbf{x}} + \Theta_i(\lambda))| < \epsilon$$

from which we get:

$$|R(g(\bar{\mathbf{x}})) - \sum_{i=1}^{N} \alpha_i f_i(\phi_h(\bar{\mathbf{x}}) + \Theta_i(\lambda))| < 2\epsilon$$

It has therefore been proven that S can approximate $R \circ g$. In order to complete the proof let us show how the approximation of the imaginary parts can be obtained.

Let us consider, for example, the approximation of $\mathbf{i}I \circ g$. By Cybenko's theorem it follows that:

$$|I(g(\bar{\mathbf{x}})) - \sum_{h=1}^{M} \beta_i \sigma(\phi_h(\bar{\mathbf{x}}) + \theta_h)| < \epsilon$$

with $\beta_h \in \Re \quad \forall h = 1, \ldots, M$

Let us consider in this case:

$$\phi_h(\bar{\mathbf{x}}) = I(\bar{\mathbf{y}}_h^T \bar{\mathbf{x}})$$

for $\bar{\mathbf{x}} \in H^n$.

Considering:

$$\theta_i(\lambda) = \lambda + \mathbf{i}\theta_i + \mathbf{j}\lambda + \mathbf{k}\lambda$$

it follows that:

$$f(\bar{\mathbf{y}}_i^T \bar{\mathbf{x}} + \Theta_i(\lambda)) = \sigma(R(\bar{\mathbf{y}}_i^T \bar{\mathbf{x}} + \lambda)) + \mathbf{i}\sigma(\phi_h(\bar{\mathbf{x}}) + \Theta_h) + \mathbf{j}\sigma(J(\bar{\mathbf{y}}_h^T \bar{\mathbf{x}} + \lambda) + \mathbf{k}\sigma(K(\bar{\mathbf{y}}_h^T \bar{\mathbf{x}} + \lambda)$$

and therefore:

$$lim_{\lambda \to -\infty} f(\bar{\mathbf{y}}_h^T \bar{\mathbf{x}} + \theta_h(\lambda)) = \mathbf{i}\sigma(\phi_h(x) + \theta_h)$$

from which:

$$lim_{\lambda \to -\infty} \sum_{h=1}^{M} \beta_h f(\bar{\mathbf{y}}_i^T \bar{\mathbf{x}} + \Theta_i(\lambda)) = \mathbf{i}\beta_h \sigma(\phi_h(x) + \theta_h)$$

Therefore a value of λ exists such that:

$$| \; \mathbf{i} I(g(\bar{\mathbf{x}})) - \sum_{h=1}^{M} \beta_h f(\bar{\mathbf{y}}_i^T \bar{\mathbf{x}} + \Theta_h(\lambda)) \; | < \epsilon$$

Remarks.
This proof represents an alternative demonstration of the density properties of CMLPs already stated [23], [44], the Complex Algebra being included in the Quaternion Algebra. Due to the identity $H = \Re^4$ the HMLP can be used to approximate continuous real valued functions of the type $f : \Re^{4n} \rightarrow \Re^4$, maintaining the advantages introduced with the HMLP in saving a number of real parameters in the structure.

5.6 Some numerical examples

5.6.1 Interpolation of 3-dimensional functions

Let us consider the electric field generated by two charges Q1=1 C and Q2=2 C, located in a 3-D space at the following positions respectively:Q1 in (0,0,0) and Q2 in (1,2,1).

The approximation of the function describing the three components of the field has been performed using both the MLP and HMLP structures. In this application the network inputs represent the coordinates of a point in the $3 - D$ space and the network outputs are the three components of the electric field in that point. This implies that the MLP has three inputs and three outputs, while the HMLP has only one input and one output neuron (the real parts of the quaternions are set to zero, i.e. the quaternionic input is of the type $q = 0 + x\mathbf{i} + y\mathbf{j} + z\mathbf{k}$ and the output is $q = 0 + E_x\mathbf{i} + E_y\mathbf{j} + E_z\mathbf{k}$).

A number of 1000 points were randomly selected in the cube with the diagonal between the points (-1,-1,-1) and (4,4,4) and the corresponding electric field values were computed. From the whole set of data, 100 patterns were used to train the networks and 900 to test the networks performance. A number of HMLPs were trained, with 3 to 8 hidden neurons. In Fig. 5.1, 5.2 and 5.3 the distribution residues of the outputs of the HMLP with 6 hidden neurons, which gave the best performance, is shown for 500 testing patterns.

A set of MLPs with 4 to 16 hidden neurons was also trained with the same number of patterns and iterations. The best results were obtained with the MLP with 14 hidden neurons. The corresponding distribution of the residues for the three real outputs is shown in Figs. 5.4, 5.5 and 5.6.

As can be observed from the figures, the performance of the HMLP and the real MLP is similar, but the number of parameters for the MLP and HMLP are very different. The 1-6-1 HMLP requires 76 real parameters while the 3-14-3 MLP requires 101. The HMLP therefore reduces the complexity of the network by about 30%.

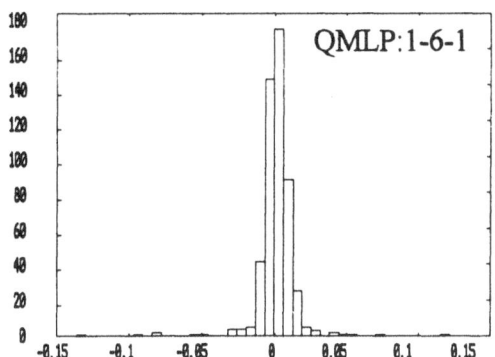

Figure 5.1: *Residue distribution of the first output of the HMLP with 6 hidden neurons*

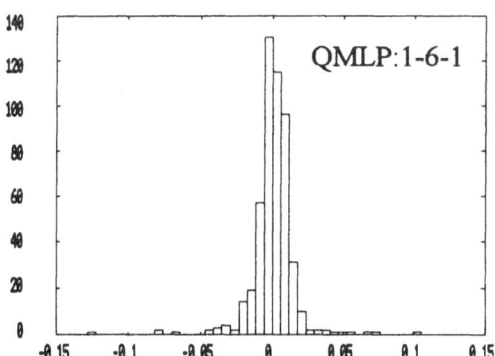

Figure 5.2: *Residue distribution of the second output of the HMLP with 6 hidden neurons*

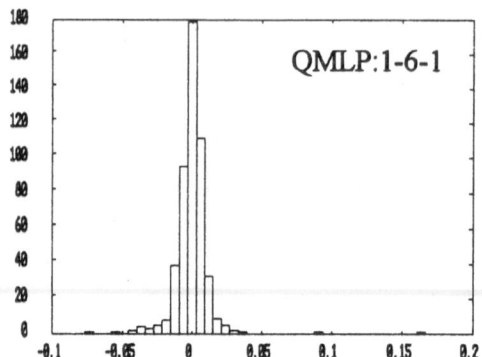

Figure 5.3: *Residue distribution of the third output of the HMLP with 6 hidden neurons*

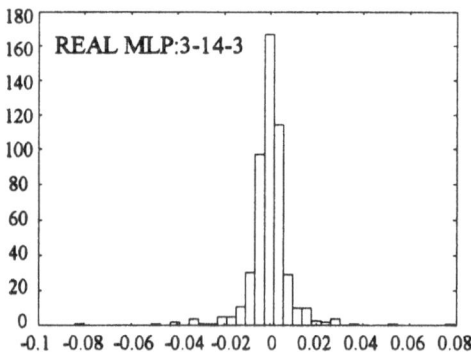

Figure 5.4: *Residue distribution for the first output of the real MLP with 14 hidden neurons*

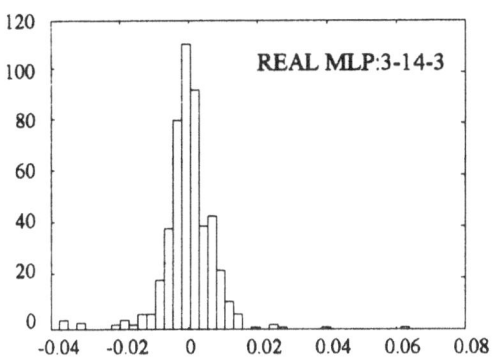

Figure 5.5: *Residue distribution for the second output of the real MLP with 14 hidden neurons*

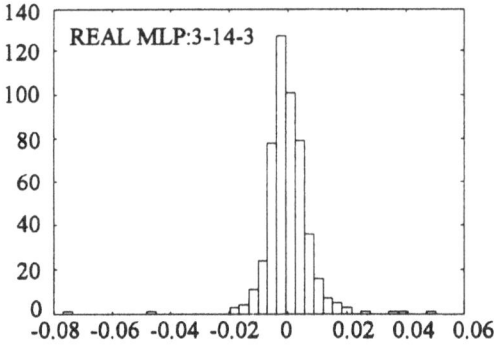

Figure 5.6: *Residue distribution for the third output of the real MLP with 14 hidden neurons*

5.6.2 A classification problem

The following example regards the classification of three species of the Iris flower: Iris setosa, Iris versicolor and Iris virginica. Classification is performed according to sepal length and width and to petal length and width. Owing to the fact that this classification involves four inputs, an HMLP with one input neuron was used. A set of 200 measures of the four inputs, together with the correct species has been used to train (50 patterns) and test (150 patterns) the network. The target were fixed as:

q=0+1i+0j+0k for Iris setosa

q=0+0i+1j+0k for Iris versicolor

q=0+0i+0j+1k for Iris virginica

Fig. 5.7 gives the sum of the square error of the output components for the testing pattern, obtained with an HMLP with one hidden neuron.

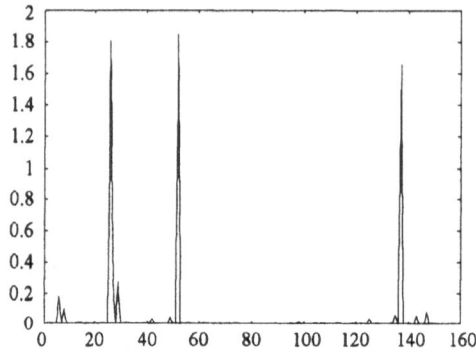

Figure 5.7: *Sum of the square error of the output components for the testing pattern, with an HMLP with one hidden neuron*

As can be observed, the HMLP only made a wrong classification on three patterns, thus showing its suitability for classification purposes.

5.7 Density theorem for the HMLP

Use of the quaternionic structures mentioned in the previous chapters should be justified from a theoretical point of view in order to assure the existence of a solution, i.e. a set of connection values able to accomplish the desired task. Such an analysis has been performed for the real MLP by several authors, [15], [19], [20], and was subsequently extended to CMLPs [22], [23], [44], as reported in the previous chapters. In this section a density theorem for continuous functions $f : X \to H$, where X is a compact subset of H^n, will be proven. As for the real MLP [15] and the CMLP [22] this result states that HMLPs with the activation function (5.1) are universal interpolators of continuous quaternion valued functions.

In order to prove the theorem the following notation is introduced:

Let $\sigma : \Re \to \Re$ be the real valued sigmoidal function:

$$\sigma(t) = \frac{1}{1 + exp(-t)}$$

and let $\mathbf{q} = q_0 + \mathbf{i}q_1 + \mathbf{j}q_2 + \mathbf{k}q_3 \in H$.

Let us consider the function $f : X \to H$:

$$f(\mathbf{q}) = \sigma(q_0) + \mathbf{i}\sigma(q_1) + \mathbf{j}\sigma(q_2) + \mathbf{k}\sigma(q_3) \qquad (5.57)$$

and the following functions from H to $\Re$:

$$R(\mathbf{q}) = q_0 \quad I(\mathbf{q}) = q_1 \quad J(\mathbf{q}) = q_2 \quad K(\mathbf{q}) = q_3$$

It follows that:

$$f = \sigma \circ R + \mathbf{i}\sigma \circ I + \mathbf{j}\sigma \circ J + \mathbf{k}\sigma \circ K$$

where $\circ$ represents the function composition operator.

Let H^n be the set of quaternion n-uples:

$$H^n = \{\bar{\mathbf{x}} = (\mathbf{x}_1, \dots, \mathbf{x}_n)^T \quad \mathbf{x}_i \in H\}$$

Having fixed $\bar{\mathbf{y}} \in H^n$, the following function can be considered:

$$\bar{\mathbf{y}}^T \cdot \bar{\mathbf{x}} = \mathbf{y}_1 \otimes \mathbf{x}_1 + \dots + \mathbf{y}_n \otimes \mathbf{x}_n$$

where the operator '$\cdot$' represents the scalar product between vectors. The function just defined is $\Re$-linear, i.e. :

$$\bar{\mathbf{y}}^T \cdot (\bar{\mathbf{x}} + \bar{\mathbf{x}}') = \bar{\mathbf{y}}^T \cdot \bar{\mathbf{x}} + \bar{\mathbf{y}}^T \cdot \bar{\mathbf{x}}'$$

$$\bar{\mathbf{y}}^T \cdot (\lambda\bar{\mathbf{x}}) = \lambda\bar{\mathbf{y}}^T \cdot \bar{\mathbf{x}} \quad \forall \lambda \in \Re$$

It is now possible to prove the following density theorem:

Theorem 5.7.1 *Let $X \subset H^n$ be a compact subset and let $g : X \to H$ be a continuous function. Then $\forall \epsilon > 0$ there exist some coefficients:*

$$\alpha_1, \ldots \alpha_n \in \Re$$

some vectors:

$$\bar{\mathbf{y}}_1, \ldots, \bar{\mathbf{y}}_N \in H^n$$

and some quaternions:

$$\Theta_1, \ldots, \Theta_N \in H$$

such that:

$$sup_{\bar{\mathbf{x}} \in X} \| \bar{g}(\bar{\mathbf{x}}) - \sum_{i=1}^{N} \alpha_i f(\bar{\mathbf{y}}_i^T \cdot \bar{\mathbf{x}} + \bar{\Theta}_i) \| < \epsilon. \qquad (5.58)$$

In other words, the real vector space:

$$S = \{ \sum_{i=1}^{N} \alpha_i f(\bar{\mathbf{y}}_i^T \cdot \bar{\mathbf{x}} + \Theta_i), \text{ where } N \text{ is a natural number, } \alpha_i \in \Re, \ \bar{\mathbf{y}}_i \in H^n, \ \Theta_i \in H \}$$

is dense in $C^0(X, H)$, the space of continuous functions $X \to H$ with the norm

$$\| g \| = sup_{\bar{\mathbf{x}} \in X} |g(\bar{\mathbf{x}})|$$

Proof:
In order to prove this theorem the following lemma is introduced:

Lemma 5.7.1 *Let $\phi : H^n \to \Re$ be a linear function.*
Then $\bar{\mathbf{y}}_R, \ \bar{\mathbf{y}}_I, \ \bar{\mathbf{y}}_J, \ \bar{\mathbf{y}}_K \in H^n$ are univocally determined such that, $\forall \bar{\mathbf{x}} \in H^n$, it follows that:

$$\phi(\bar{\mathbf{x}}) = R(\bar{\mathbf{y}}_R^T \cdot \bar{\mathbf{x}}) = I(\bar{\mathbf{y}}_I^T \cdot \bar{\mathbf{x}}) = J(\bar{\mathbf{y}}_J^T \cdot \bar{\mathbf{x}}) = K(\bar{\mathbf{y}}_K^T \cdot \bar{\mathbf{x}})$$

Proof of Lemma 5.7.1
The proof is only given for the representation $\phi(\bar{\mathbf{x}}) = R(\bar{\mathbf{y}}_R^T \cdot \bar{\mathbf{x}})$, as it is easy to extend it to all other cases.
Let us define the following application:

$$\Phi : H^n \to \{\phi : H^n \to \Re : \phi \text{ is linear}\}$$

defined as: $\Phi(\mathbf{y}) = \phi$ such that $\phi(\bar{\mathbf{x}}) = R(\bar{\mathbf{y}}^T \cdot \bar{\mathbf{x}}) \ \forall \bar{\mathbf{x}} \in H^n$.
It has to be proven that Φ is a one-to-one correspondence.
It should be observed that Φ is a linear application between vector spaces of the same dimension. To prove the lemma it is therefore sufficient to show that Φ is injective.
This means that if $\bar{\mathbf{y}} \in H^n$ is such that $R(\bar{\mathbf{y}}_R^T \cdot \bar{\mathbf{x}}) = 0 \quad \forall \bar{\mathbf{x}} \in H^n$, it follows that $\bar{\mathbf{y}} = 0$.

Let us prove what has been stated.
Let us consider $\bar{\mathbf{y}} = (\mathbf{y}_1, \ldots, \mathbf{y}_n)^T$ and take into account, for the sake of simplicity, the first component $\mathbf{y}_1 \in H$.
Owing to the fact that:

$$0 = R(\bar{\mathbf{y}}^T \cdot \mathbf{e}_1) = R(\bar{\mathbf{y}}^T \cdot (\mathbf{i}\mathbf{e}_1)) = R(\bar{\mathbf{y}}^T \cdot (\mathbf{j}\mathbf{e}_1)) = R(\bar{\mathbf{y}}^T \cdot (\mathbf{k}\mathbf{e}_1))$$

if $\mathbf{e}_1 = (1\ 0\ \ldots\ 0)^T$ it follows that:

$$R(\mathbf{y}_1) = I(\mathbf{y}_1) = J(\mathbf{y}_1) = K(\mathbf{y}_1) = 0 \quad \rightarrow \quad \mathbf{y}_1 = 0$$

In the same way it can be shown that $\mathbf{y}_i = 0$ when $i = 2, \ldots, n$ and therefore $\bar{\mathbf{y}} = 0$
The lemma has therefore been proven.
Proof of Theorem 5.7.1
Let us consider the following continuous function $g : X \rightarrow H$ together with its four components:

$$g = g_0 + \mathbf{i}g_1 + \mathbf{j}g_2 + \mathbf{k}g_3 = R \circ g + \mathbf{i}I \circ g + \mathbf{j}J \circ g + \mathbf{k}K \circ g$$

If S can separately approximate the functions:

$$R \circ g, \quad \mathbf{i}I \circ g, \quad \mathbf{j}J \circ g, \quad \mathbf{k}K \circ g$$

then of course S can approximate g.
Let us first consider the real part of g, denoted as $R \circ g : X \rightarrow \Re$.
From Cybenko's density theorem in the real field applied to $C^0(X, \Re)$, (where X can be seen as a compact subset of $\Re^{4n}$) it follows that there exist some real coefficients $a_1, \ldots, a_N$, some sigmoidal real valued functions $\phi_1, \ldots \phi_N$ on $\Re^{4n} = H^n$ and some real parameters $\theta_1, \ldots, \theta_N$ such that:

$$\left| R(g(\bar{\mathbf{x}})) - \sum_{i=1}^{N} a_i \sigma(\phi_i(\bar{\mathbf{x}}) + \theta_i) \right| < \epsilon$$

$for\ all\ \bar{\mathbf{x}} \in X$.
For the lemma already proven we have $\phi_i(\bar{\mathbf{x}}) = R(\bar{\mathbf{y}}_i^T \cdot \bar{\mathbf{x}})$ (where $\bar{\mathbf{x}}$ is now a vector of H^n and $\bar{\mathbf{y}}_i \in H^n$).
Let us consider:

$$\Theta_i(\lambda) = \theta_i + \mathbf{i}\lambda + \mathbf{j}\lambda + \mathbf{k}\lambda$$

with $\lambda \in \Re$.
The function f can therefore be represented as:
$$f(\bar{\mathbf{y}}_i^T \cdot \bar{\mathbf{x}} + \Theta_i(\lambda)) = \sigma(\phi_i(\bar{\mathbf{x}}) + \Theta_i) + \mathbf{i}I(\bar{\mathbf{y}}_i^T \cdot \bar{\mathbf{x}} + \lambda) + \mathbf{j}J(\bar{\mathbf{y}}_i^T \cdot \bar{\mathbf{x}} + \lambda) + \mathbf{k}K(\bar{\mathbf{y}}_i^T \cdot \bar{\mathbf{x}} + \lambda)$$
Since $\bar{\mathbf{x}}$ is in the compact set X and $\bar{\mathbf{y}}_i$ are fixed, the imaginary parts of the product $\bar{\mathbf{y}}_i^T \cdot \bar{\mathbf{x}}$ are bounded.

Then, if $\lambda \to -\infty$ the imaginary parts of the function $f(\bar{\mathbf{y}}_i^T \cdot \bar{\mathbf{x}} + \Theta_i(\lambda))$ converge uniformly to 0 and the following convergence:

$$lim_{\lambda \to -\infty} f(\bar{\mathbf{y}}_i^T \cdot \bar{\mathbf{x}} + \Theta_i(\lambda)) = \sigma(\phi_i(\bar{\mathbf{x}}) + \Theta_i)$$

is uniform, for all $i = 1, \ldots, N$.
It follows that a value of λ exists such that:

$$|\sum_{i=1}^{N} \alpha_i \sigma(\phi_i(\bar{\mathbf{x}}) + \theta_i) - \sum_{i=1}^{N} \alpha_i f_i(\bar{\mathbf{y}}_i^T \bar{\mathbf{x}} + \Theta_i(\lambda))| < \epsilon$$

from which we get:

$$|R(g(\bar{\mathbf{x}})) - \sum_{i=1}^{N} \alpha_i f_i(\phi_h(\bar{\mathbf{x}}) + \Theta_i(\lambda))| < 2\epsilon$$

It has therefore been proven that S can approximate $R \circ g$. In order to complete the proof let us show how the approximation of the imaginary parts can be obtained.
Let us consider, for example, the approximation of $i I \circ g$. By Cybenko's theorem it follows that:

$$|I(g(\bar{\mathbf{x}})) - \sum_{h=1}^{M} \beta_i \sigma(\phi_h(\bar{\mathbf{x}}) + \theta_h)| < \epsilon$$

with $\beta_h \in \Re \quad \forall h = 1, \ldots, M$
Let us consider in this case:

$$\phi_h(\bar{\mathbf{x}}) = I(\bar{\mathbf{y}}_h^T \bar{\mathbf{x}})$$

for $\bar{\mathbf{x}} \in H^n$.
Considering:

$$\theta_i(\lambda) = \lambda + i\theta_i + j\lambda + k\lambda$$

it follows that:

$$f(\bar{\mathbf{y}}_i^T \bar{\mathbf{x}} + \Theta_i(\lambda)) = \sigma(R(\bar{\mathbf{y}}_i^T \bar{\mathbf{x}} + \lambda)) + i\sigma(\phi_h(\bar{\mathbf{x}}) + \Theta_h) + j\sigma(J(\bar{\mathbf{y}}_h^T \bar{\mathbf{x}} + \lambda) + k\sigma(K(\bar{\mathbf{y}}_h^T \bar{\mathbf{x}} + \lambda)$$

and therefore:

$$lim_{\lambda \to -\infty} f(\bar{\mathbf{y}}_h^T \bar{\mathbf{x}} + \theta_h(\lambda)) = i\sigma(\phi_h(x) + \theta_h)$$

from which:

$$lim_{\lambda \to -\infty} \sum_{h=1}^{M} \beta_h f(\bar{\mathbf{y}}_i^T \bar{\mathbf{x}} + \Theta_i(\lambda)) = i\beta_h \sigma(\phi_h(x) + \theta_h)$$

Therefore a value of λ exists such that:

$$| \, \mathbf{i} I(g(\bar{\mathbf{x}})) - \sum_{h=1}^{M} \beta_h f(\bar{\mathbf{y}}_i^T \bar{\mathbf{x}} + \Theta_h(\lambda)) \, | < \epsilon$$

Remarks.
This proof represents an alternative demonstration of the density properties of CMLPs already stated [23], [44], the Complex Algebra being included in the Quaternion Algebra. Due to the identity $H = \Re^4$ the HMLP can be used to approximate continuous real valued functions of the type $f : \Re^{4n} \to \Re^4$, maintaining the advantages introduced with the HMLP in saving a number of real parameters in the structure.

5.8   Some numerical examples

5.8.1   Interpolation of 3-dimensional functions

Let us consider the electric field generated by two charges Q1=1 C and Q2=2 C, located in a 3-D space at the following positions respectively:Q1 in (0,0,0) and Q2 in (1,2,1).
The approximation of the function describing the three components of the field has been performed using both the MLP and HMLP structures. In this application the network inputs represent the coordinates of a point in the $3-D$ space and the network outputs are the three components of the electric field in that point. This implies that the MLP has three inputs and three outputs, while the HMLP has only one input and one output neuron (the real parts of the quaternions are set to zero, i.e. the quaternionic input is of the type $q = 0 + x\mathbf{i} + y\mathbf{j} + z\mathbf{k}$ and the output is $q = 0 + E_x\mathbf{i} + E_y\mathbf{j} + E_z\mathbf{k}$).
A number of 1000 points were randomly selected in the cube with the diagonal between the points (-1,-1,-1) and (4,4,4) and the corresponding electric field values were computed. From the whole set of data, 100 patterns were used to train the networks and 900 to test the networks performance. A number of HMLPs were trained, with 3 to 8 hidden neurons. In Fig. 5.1, 5.2 and 5.3 the distribution residues of the outputs of the HMLP with 6 hidden neurons, which gave the best performance, is shown for 500 testing patterns.
A set of MLPs with 4 to 16 hidden neurons was also trained with the same number of patterns and iterations. The best results were obtained with the MLP with 14 hidden neurons. The corresponding distribution of the residues for the three real outputs is shown in Figs. 5.4, 5.5 and 5.6.

As can be observed from the figures, the performance of the HMLP and the real MLP is similar, but the number of parameters for the MLP and HMLP are very different. The 1-6-1 HMLP requires 76 real parameters while the 3-14-3 MLP requires 101. The HMLP therefore reduces the complexity of the network by about 30%.

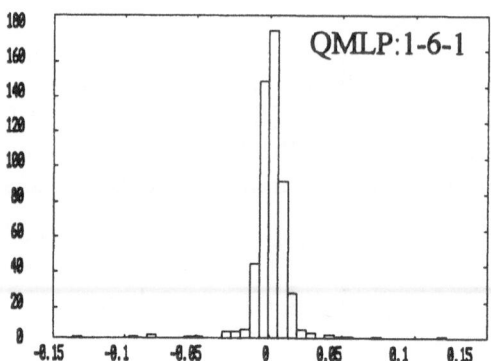

Figure 5.8: *Residue distribution of the first output of the HMLP with 6 hidden neurons*

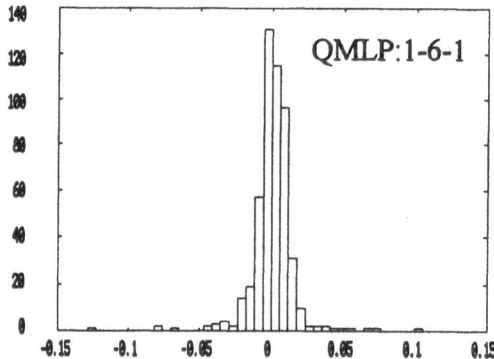

Figure 5.9: *Residue distribution of the second output of the HMLP with 6 hidden neurons*

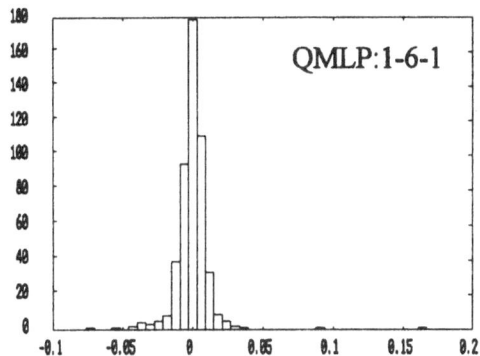

Figure 5.10: *Residue distribution of the third output of the HMLP with 6 hidden neurons*

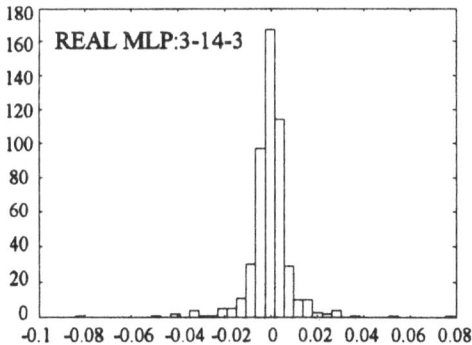

Figure 5.11: *Residue distribution for the first output of the real MLP with 14 hidden neurons*

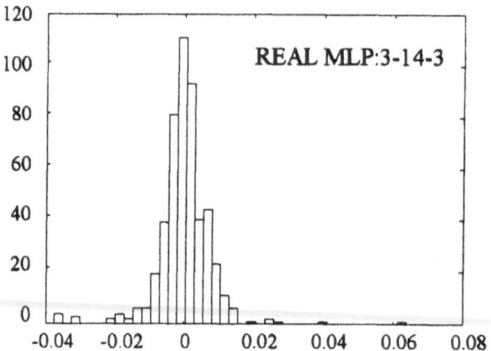

Figure 5.12: *Residue distribution for the second output of the real MLP with 14 hidden neurons*

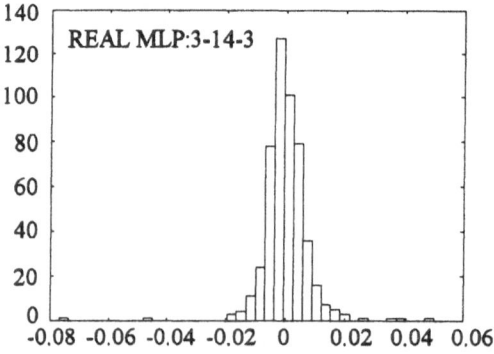

Figure 5.13: *Residue distribution for the third output of the real MLP with 14 hidden neurons*

5.8.2   A classification problem

The following example regards the classification of three species of the Iris flower: Iris setosa, Iris versicolor and Iris virginica. Classification is performed according to sepal length and width and to petal length and width. Owing to the fact that this classification involves four inputs, an HMLP with one input neuron was used. A set of 200 measures of the four inputs, together with the correct species has been used to train (50 patterns) and test (150 patterns) the network. The target were fixed as:

q=0+1i+0j+0k for Iris setosa

q=0+0i+1j+0k for Iris versicolor

q=0+0i+0j+1k for Iris virginica

Fig. 5.7 gives the sum of the square error of the output components for the testing pattern, obtained with an HMLP with one hidden neuron.

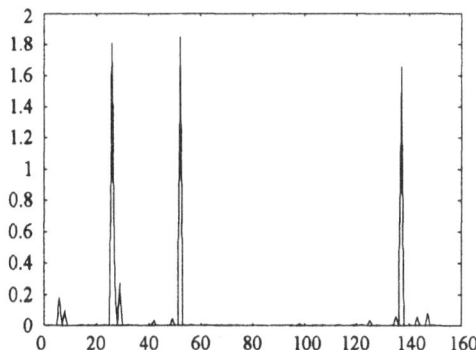

Figure 5.14: *Sum of the square error of the output components for the testing pattern, with an HMLP with one hidden neuron*

As can be observed, the HMLP only made a wrong classification on three patterns, thus showing its suitability for classification purposes.

Chapter 6

Chaotic time series prediction with CMLP and HMLP

6.1 Introduction

In the preceding chapters the approximation capabilities of CMLPs and HMLPs were outlined, proving a number of theoretical results. This chapter is mainly devoted to showing this approximation performance with reference to the prediction of chaotic, complex valued time series characterizing light transmission through non-linear dielectric materials. HMLPs will also be applied to the estimation of hypercomplex time series generated by well-known circuits such as Chua's circuit and others.

6.2 Modelling of non-linear systems

The problem of modelling and predicting chaotic time series is a very interesting one in various scientific fields. It has several points in common with dynamic system modelling; in fact both problems use the same methodologies when nothing is known a priori about the system dynamics but a number of input-output measures.

The inability to predict certain very complicated time series up to a few years ago was attributed to the lack of truly efficient estimation algorithms. Recently the discovery of chaos and its characteristics has solved such problems, showing that the unpredictability of steady-state motions of chaotic systems is indeed a 'structural' property, due to 'sensitive dependence on initial conditions' [43]. The problem of chaotic time series determination therefore has to be reformulated in terms of short-term prediction, owing to the degradation in the prediction accuracy as the prediction step P increases.

In the last few years neural networks, and in particular Multi-layer Perceptrons (MLP) [13], have been used to perform non-linear function approximation as well as to predict non-linear time series [44]. In particular, as regards chaotic time series, where only short-term prediction can be obtained, great accuracy is required when the prediction time increases. In this field, neural networks allow better results than other strategies [44], due to their ability to extract the non-linear maps generating the data.

However, when the number of I/O variables grows, a drawback arises, due to the increasing number of connections in the MLP, i.e. of free parameters in the learning algorithm. This fact degrades network performance, increasing the possibility of being trapped in local minima and requiring a great number of I/O samples during learning.

To alleviate this problem new neural structures, developed in both complex and quaternion algebra can be used, when the signals involved are complex or quaternionic, respectively. As will be shown in the following sections, such structures provide the same approximation capabilities as the real MLP with a lower number of free parameters, thus improving the convergence of the learning algorithm.

6.2.1 Neural networks for non-linear systems prediction

The problem of system dynamics prediction can be addressed as a task of deriving the relations which, given the input values at time t, allow the corresponding outputs to be determined as a function of the input values and a suitable number of delayed output values. Such a relation is in general derived starting from a set of measured input-output samples. For SISO (Single-Input, Single-Output) systems, the problem consists of obtaining the relation, which in general is a non-linear one, between the output $y(t)$ and the vector

$$\phi(t) = [y(t-1)\ldots y(t-n_a)u(t-1)\ldots u(t-n_b)] \qquad (6.1)$$

where $t = 1, 2, \ldots$ and $u(\cdot)$ is the input.

The number of delayed input-output values is in general unknown and should be determined using *trial and error* techniques, with the help of some generalization of the strategies used in linear system theory. In order to determine such a relation, a set of N measured I/O samples has to be available. In neural identification techniques, this set constitutes the learning set:

$$Z^N = [y(t), \phi(t)] \quad t = 1, \ldots, N \qquad (6.2)$$

To approach the problem, a certain class of models, i.e. of I/O relations, has to be considered. If the system is fully unknown, (black-box identification), this class should be as wide as possible. In the case of non-linear systems, one possibility is to consider the input-output relations in the class of functions which can be parametrized by an MLP, due to the density results reported previously [41]. This means that the unknown function is approximated as a linear combination of a finite number g_k of sigmoidal basis functions, as follows:

$$\hat{y}(t) = \sum_{k=1}^{n} \theta(k) g_k(\phi(t)) \qquad (6.3)$$

The parameters θ should minimize a quadratic performance index which represents the error between the measured and the actual values of the outputs. In the neural identification framework the network weights and biases represent the model parameters and the learning algorithm stands for the identification algorithm. If the measured data are assumed to be obtained by:

$$y(t) = g_0(\phi(t)) + e(t) \qquad (6.4)$$

where $e(t)$ is a white noise with variance λ, it can be shown that the expected value of the squared error between the system and the model output is proportional to $\lambda m/N$, where m is the number of parameters in the model [40]. Each strategy able to perform the identification task adopting a lower number of parameters should therefore be considered. Among these strategies, when handling MIMO systems, the CMLP or the HMLP could be used to collapse a number of input variables into one neuron and therefore reduce the network complexity. However, in neural identification all the validation techniques commonly used in non linear system identification should also be used. Moreover, particular attention should be devoted to the problems encountered while using neural networks, namely, determining the correct topology, selecting a complete set of learning patterns and avoiding the problem of overlearning.

6.2.2   Chaotic time series prediction

A simple and intuitive definition of a chaotic system is that of a deterministic system which shows random behaviour [43].

A more rigorous definition is reported in the appendix, where the fundamental features of chaos, consisting in the strong dependence of the dynamic on the initial conditions, is formally defined. Such a characteristic makes a chaotic system unpredictable at large time steps. The problem of predicting a sequence of data generated by a chaotic system (chaotic time series) can therefore only be formulated as a short-term prediction problem, because the prediction accuracy strongly degrades at high prediction steps, regardless of the prediction strategy adopted.

The use of neural networks as prediction tools for chaotic systems has shown that they are much more accurate than other methods, in particular at large prediction steps [44], [45]. This feature depends on the ability of neural networks to extract the non-linear map which generated the data, starting from a set of input-output samples. To perform chaotic time series prediction with a neural approach, a neural network is trained to give the estimated value of the time series τ steps ahead with respect to the input. The desired input-output relation which should be approximated by the MLP is therefore:

$$\bar{x}(t + \tau) = f(\bar{x}(t)) \tag{6.5}$$

where $\bar{x}$ is the vector of the state variables of the system at time t. To perform the prediction, two different strategies can be adopted. The first strategy is called the 'direct method' and consists of fixing the prediction step τ and training the network to predict the value $\bar{x}(t + \tau)$ when the measured value $\bar{x}(t)$ is impinged to the input layer. The second strategy, called the 'iterative method', consists of fixing a lower prediction step, for example $\tau = 1$, and feeding the network output $\bar{x}(t + \tau)$ back to the input layer for a fixed number n of time steps. After n time steps, the measured value $\bar{x}(t + n\tau)$ is presented to the input layer. Even if the second strategy seems less accurate, due to the amplification of the estimation error caused by the feedback, it has been shown that this strategy gives better performance at large prediction steps. The error introduced by the feedback is in fact compensated for by the greater accuracy of the estimation when $\tau = 1$ [44]. In the following applications the iterative method will be adopted.

When chaotic systems are characterized by two, three or four state variables, the CMLP or the HMLP can be used for the prediction. In order to show the improvement obtained with the HMLP structure, in all the applications, a comparison will be made with the performance obtained with the real MLP. Due to the chaotic nature of the systems considered, in order to evaluate the prediction accuracy, a suitable performance index is computed on a large number of samples not considered during the learning phase. This performance index, together with the usual quadratic output error, gives further guarantees of the prediction ability of the neural network, and is commonly adopted in chaotic time series prediction problems [46], [24]. The performance index is the following:

$$\rho_s(\tau) = \frac{\displaystyle\sum_{t=1}^{M}(O_s(t) - O_{sm})(O'_s(t) - O'_{sm})}{\sqrt{\displaystyle\sum_{t=1}^{M}(O_s(t) - O_{sm})^2}\sqrt{\displaystyle\sum_{t=1}^{M}(O'_s(t) - O'_{sm})^2}} \tag{6.6}$$

where τ denotes the prediction step, M the number of samples used for the testing phase, $O_s(t)$ the s-th τ steps ahead output component of the generic neuron of the output layer, evaluated for the t-th time series element $t = 1, \ldots, M$, $O_{sm}(t)$ is the mean value of $O_s(t)$ on the whole set of testing patterns, and $O'_s(t)$ and $O'_{sm}(t)$ are the corresponding terms of the measured time series. In particular, $s = 0, 1$ for the complex time series, and $s = 0, 1, 2, 3$ for the quaternionic one. Correlation index values close to 1 for a particular prediction step τ mean that the τ-step ahead zero-mean predicted time series is close to the measured one on the whole set of samples available for the testing phase. In order to make a better comparison between the approximation performance of the networks, each training phase consisted of the same number of learning cycles, with the same fixed values for the learning parameters.

In the following section an application of the CMLP to the prediction of a chaotic time series is outlined.

6.3   Chaotic phenomena in light transmission

The light transmitted from a ring cavity containing a non-linear dielectric medium undergoes transitions from a stationary state to periodic and non-periodic states when the intensity of the incident light is increased. The non-periodic state is characterized by a chaotic variation of the light intensity and associated broadband noise in the power spectrum. The physical phenomenon under consideration has been paid much attention due to its applicability as an optical device. Let us consider a ring cavity containing a non-linear dielectric medium of length l, as illustrated in Fig. 6.1. Mirrors 1 and 2 have reflectivity R, while mirrors 3 and 4 have 100% reflectivity, so that a part of the light transmitted is fed back to the medium. Assuming that the response of the medium is described by the Debye relaxation equation, the Maxwell-Debye equations which govern the dynamics of the system, integrated with respect to the space variable and suitably sampled, lead to the following complex time series, also known as the *Ikeda Map* [24]:

$$E_x(t+1) = a + b(E_x(t)cos(z(t)) - E_y(t)sin(z(t))) \qquad (6.7a)$$
$$E_y(t+1) = b(E_x(t)sin(z(t)) + E_y(t)sin(z(t))) \qquad (6.7b)$$

in which:
$$z(t) = 0.4 - \frac{6.0}{1 + E_x^2(t) + E_y^2(t)}$$

$$a = 1, \quad b = 0.7, \quad E_x(0) = E_y(0) = 0, \qquad (6.8)$$

where : E_x, E_y respectively are the real and imaginary parts of the electric field in the cavity. The parameter a is proportional to the amplitude of the incident field, while parameter b characterizes the dissipation of the electric field in the cavity. Under the conditions and parameter values reported in (6.8), the complex time series (6.7) takes on chaotic motions.

6.3.1   Estimation of the Ikeda Map via CMLPs and MLPs

A chaotic time series like the *Ikeda Map* is a suitable benchmark to evaluate the estimation capabilities of the CMLP. In this paragraph, both a real and a complex MLP will be considered. The comparison will show that the use of CMLPs is advantageous from several points of view [6]. All the topologies considered are built with only one hidden layer, owing to the density theorem reported in [15]. The real MLP structure needs two input and two output neurons in order to predict the complex behavior of the time series (6.7), while the complex MLP requires one input and one output unit. With the initial conditions reported in (6.8) a set of 2000 patterns was computed simulating the time series (6.7). A subset of 100 patterns was used in order to perform network training, while the remaining 1900 were used in the testing phase. Moreover, as the strategies for topological optimization have not been deeply investigated, the suitable number of hidden units was chosen, both for the real and the complex MLPs, with the so called 'growing method', i.e. starting with a low number of hidden neurons and increasing the structure until suitable performance is reached as regards the generalization characteristics after a fixed number of learning cycles, for all the topologies, equal to $M_{max} = 20000$. As regards the generalization capabilities, in order to evaluate the prediction results, the correlation $\rho(\tau)$ as defined previously, is adopted. The growing strategy was first adopted for the complex MLP: Figs. 6.2 and 6.3 show the trends of the correlation index for different complex MLP topologies with a number of hidden units growing from 3 to 8. In particular Fig.6.2 shows the imaginary part of the Ikeda map, while Fig.6.3 shows the real part.

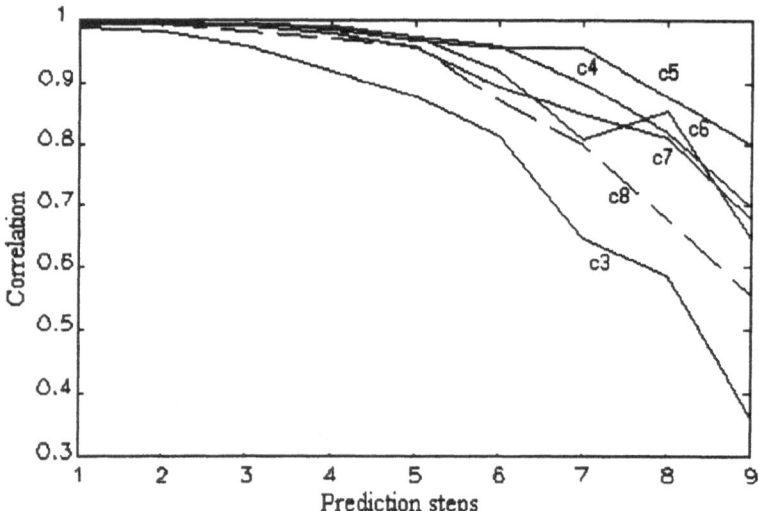

Figure 6.1: *Prediction of the imaginary part of the Ikeda map with CMLPs*

It can be observed that the best performance during the testing phase with 1900 patterns was reached with a topology containing 5 complex hidden neurons. The same technique was applied to real MLP. Fig.6.4 and Fig.6.5 report

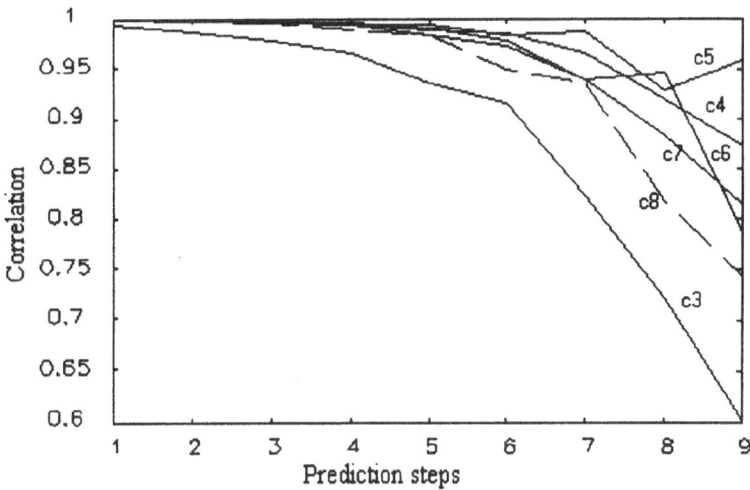

Figure 6.2: *Prediction of the real part of the Ikeda map with CMLPs*

the correlation trends for the real MLP structures growing from 8 to 16 neurons in the hidden layer.

These results show that a neural networks with 12 neurons perform better than all the other real MLPs trained. Figs. 6.6 and 6.7 compare the correlation trends of a real MLPs with 12 neurons and a complex MLP with 5 hidden units.   From these figures it derives that a complex MLP with 5 hidden units has a correlation trend better than that of a real MLP with 12 hidden units. It should also be pointed out that the real MLP requires 62 real parameters to perform the time series estimation, while the complex one employs 32 real parameters. The suitability of the complex MLP in complex time series estimation is therefore evident. Moreover, it has to be observed that all the topologies employed behave better than the results reported in [63] where the correlation index τ, obtained with a RBF (radial basis function) neural network with 10 hidden units, reaches the value of 0.1 when P=6.

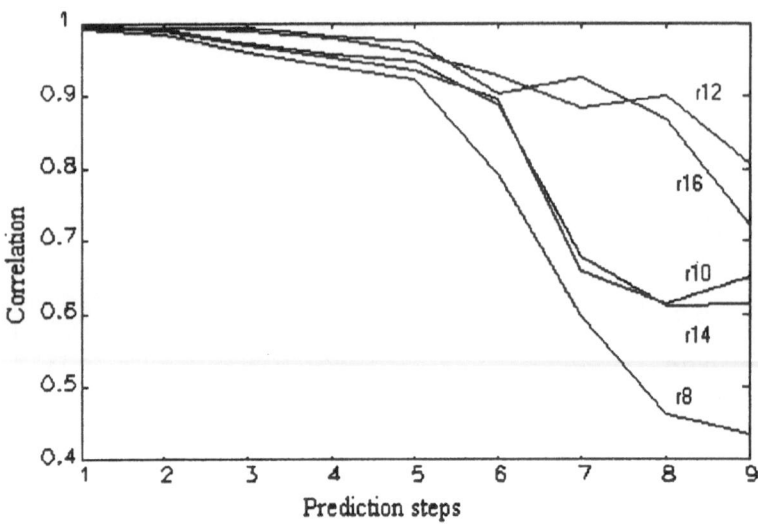

Figure 6.3: *Prediction of the imaginary part of the Ikeda map with MLPs*

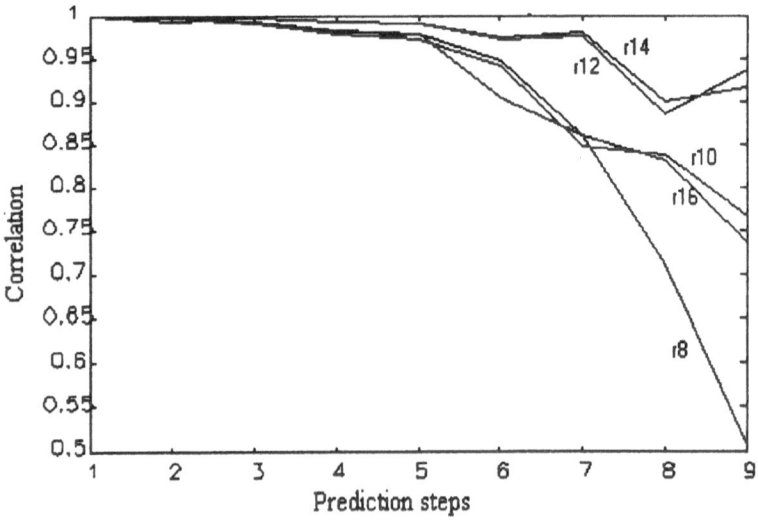

Figure 6.4: *Prediction of the real part of the Ikeda map with MLPs*

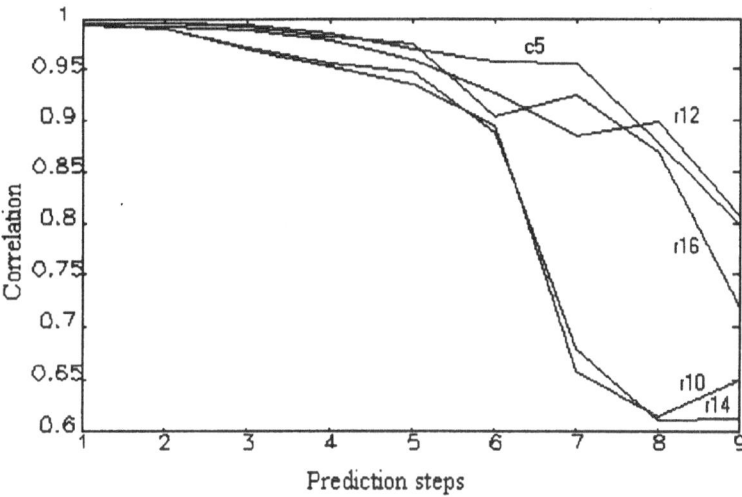

Figure 6.5: *Comparison between MLP and CMLP on prediction of the imaginary part of the Ikeda map*

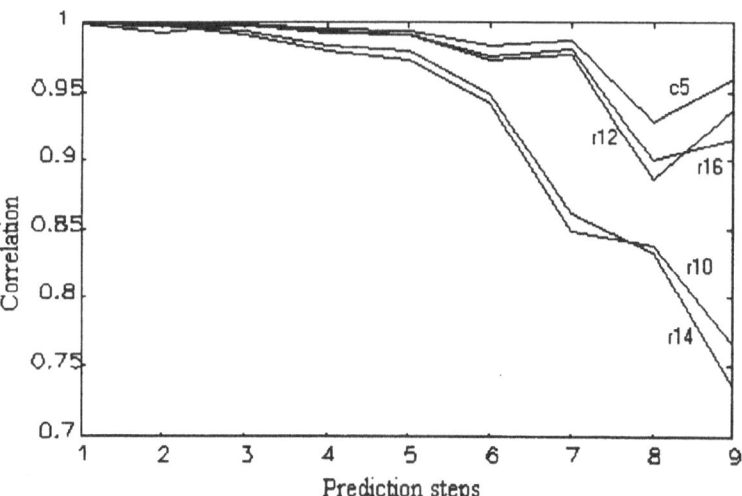

Figure 6.6: *Comparison between MLP and CMLP on prediction of the real part of the Ikeda map*

6.4   Hypercomplex chaotic time series

In this section, further examples of chaotic time series prediction involving three
or four input variables are reported, and the HMLP has therefore been adopted.
The HMLP structure is used to perform short-term prediction of the time series
generated by some canonical chaotic systems: Chua's circuit [43], [48], a Saito
circuit [49] and the Lorenz system [47]. The correlation index described will be
evaluated for a set of neural topologies, both real and hypercomplex, and for
different values of τ, in order to select the network with the best performance.

6.4.1   Chua's circuit

Chua's circuit is known as the simplest autonomous circuit showing a large set
of chaotic behaviour [43]. Its state equations are as follows:

$$\dot{x} = \alpha(y - h(x)) \tag{6.9a}$$

$$\dot{y} = x - y + z \tag{6.9b}$$

$$\dot{z} = \beta y - \gamma z \tag{6.9c}$$

where:

$$h(x) = m_1 x + 0.5(m_0 - m_1)(\mid x + 1 \mid - \mid x - 1 \mid)$$

The attractor known as 'double-scroll' [43] is obtained with the parameters

$$(\alpha, \beta, \gamma, m_0, m_1) = (9, 14.286, 0, -1/7, 2/7).$$

The corresponding time series was obtained by simulating the system with
$\Delta T = 0.02sec$ and the initial condition $(0.1, 0.1, 0.1)$. A set of 250 terms of the
simulated time series were used to train the neural structures. As regards the
HMLP, only one input and one output neuron are needed, whose values are the
quaternions $q(t) = 0 + ix(t) + jy(t) + kz(t)$ and $q(t+1) = 0 + ix(t+1) + jy(t+1) + kz(t+1)$ respectively. This means that in this case vector quaternions (i.e.
quaternions with the real part equal to zero) are used; the I/0 space dimension
is, in fact, three. The corresponding real MLP has 3 input and 3 output
neurons. Several topologies were trained for a fixed number of learning cycles
to compare their performance. The best results, in terms of correlation values
for 250 I/O samples not used during the training phase, were obtained with 12
hidden units for the real MLP and 3 hidden units for the HMLP. In Fig.6.8,
6.9 and 6.10 a comparison is made between the correlation values of the state
variables versus the prediction time steps for both the MLP and the HMLP.
As can be observed, the HMLP leads to better correlation values than the real
MLP. Moreover the 1-3-1 HMLP has 40 real parameters (including the biases)
while the 3-12-3 MLP requires 87 parameters. The improvement introduced by
the HMLP is evident. Fig.6.11-Fig.6.13 show the five-step-ahead predictions
for the three components of the output of the $1 - 3 - 1$ HMLP.   Fig.6.14

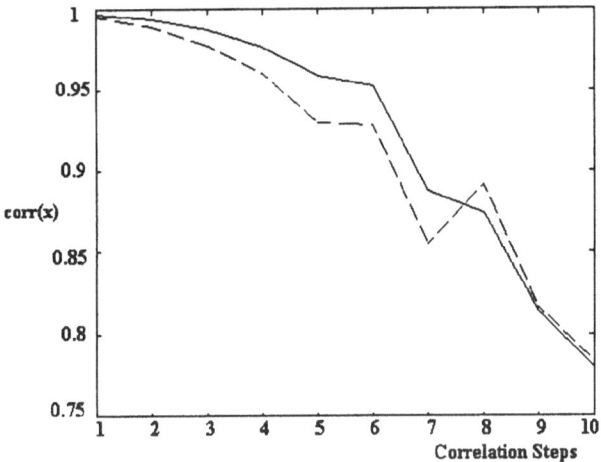

Figure 6.7: *Comparison between the correlation trends of the first output of the 3 − 12 − 3 MLP (- -) and the first imaginary component of the 1 − 3 − 1 MLP (-) versus the prediction steps*

and Fig.6.15 show the simulated and the five-step-ahead predicted attractors of Chua's circuit computed on 400 testing patterns.

As can be observed from these figures, the prediction accuracy is appreciable. Therefore, although the approximation capabilities of the real MLP and the HMLP have been demonstrated to be the same, in several cases, like this one, the results obtained with the HMLP are much better than those obtained with the other structure. This may be due to the lower number of real parameters involved in the optimization algorithm, which leads to a simpler and faster learning phase.

6.4.2   Saito's Circuit

Saito's circuit, introduced in [49], is characterized by 4 state variables and five parameters which determine transition from torus doubling route to area and volume expanding chaos. The circuit dynamics is governed by the following equations:

$$
\begin{bmatrix} \dot{x}_1 \\ \dot{y}_1 \end{bmatrix} = \begin{bmatrix} -1 & 1 \\ -\alpha_1 & \alpha_1\beta_1 \end{bmatrix} \begin{bmatrix} x_1 - \eta p_1 h(z) \\ y_1 - \eta \frac{p_1}{\beta_1} h(z) \end{bmatrix} \tag{6.10}
$$

$$
\begin{bmatrix} \dot{x}_2 \\ \dot{y}_2 \end{bmatrix} = \begin{bmatrix} -1 & 1 \\ -\alpha_2 & \alpha_2\beta_2 \end{bmatrix} \begin{bmatrix} x_2 - \eta p_2 h(z) \\ y_2 - \eta \frac{p_2}{\beta_2} h(z) \end{bmatrix} \tag{6.11}
$$

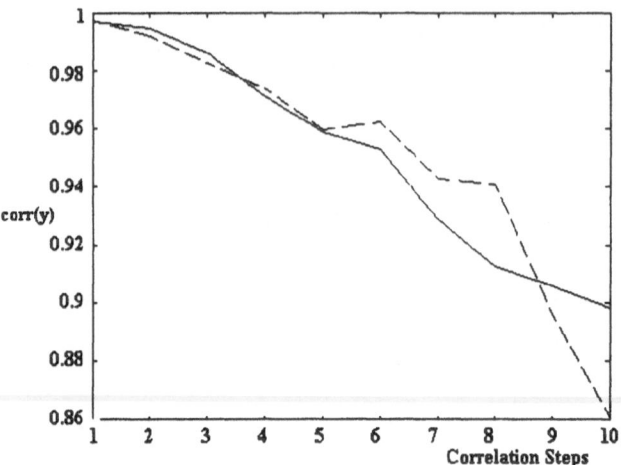

Figure 6.8: *Comparison between the correlation trends of the second output of the 3 − 12 − 3 MLP (- -) and the second imaginary component of the 1 − 3 − 1 HMLP (-) versus the prediction steps*

where:

$$h(z) = \begin{cases} 1 & \text{for } z \geq -1 \\ -1 & \text{for } z \leq 1 \end{cases}$$

and $z = x_1 + x_2$, $p_1 = \dfrac{\beta_1}{1 - \beta_1}$, $p_2 = \dfrac{\beta_2}{1 - \beta_2}$

The system is characterized by two positive Lyapunov exponents which make the circuit behaviour hyperchaotic. The corresponding time series was obtained by simulating the system with

$$(\alpha_1, \beta_1, \alpha_2, \beta_2, \eta) = (7.5,\ 0.16,\ 15,\ 0.097,\ 1.3),$$

starting from the initial condition

$$(x_1, y_1, x_2, y_2) = (1, 0, 1, 0)$$

and with $\Delta T = 0.002 sec$.

As in the previous application, several neural topologies were trained with 250 I/O samples. In Fig.6.16-6.19 a comparison is made between the correlation trends obtained with the 1-2-1 HMLP and the 4-8-4 MLP for the state variable.

Fig.6.20 and Fig.6.21 show the simulated and the five-step-ahead predicted attractors of Saito's circuit computed on the testing patterns.

As it can be observed, here again the HMLP performs better at large prediction steps. The correlation index for the other state variables shows a similar trend. The number of free parameters is 28 and 76 respectively.

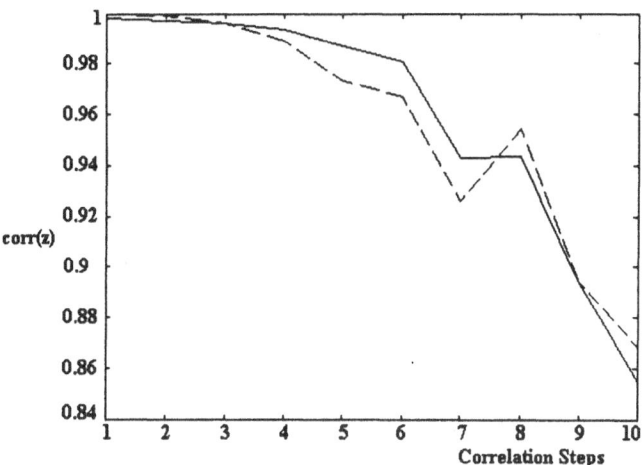

Figure 6.9: *Comparison between the correlation trends of the third output of the $3 - 12 - 3$ MLP (- -) and the third imaginary component of the $1 - 3 - 1$ MLP (-) versus the prediction steps*

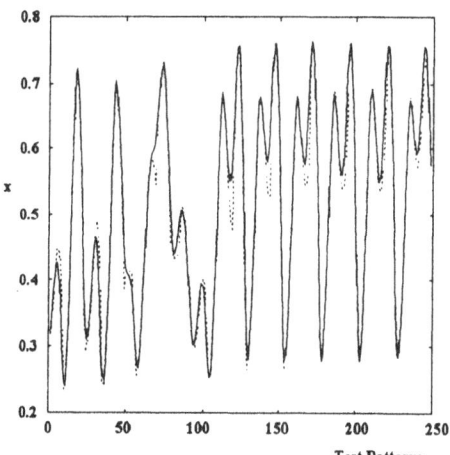

Figure 6.10: *Comparison between the trends of the first imaginary component of the $1 - 3 - 1$ HMLP (-) and the corresponding target versus the number of testing patterns*

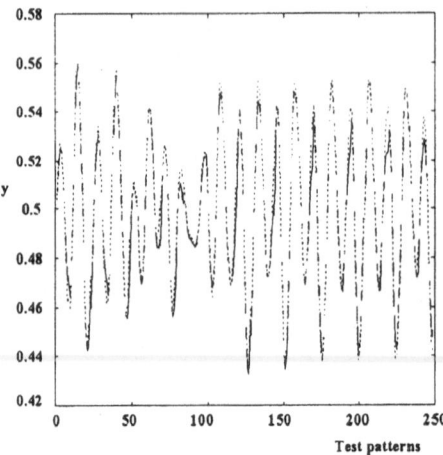

Figure 6.11: *Comparison between the trends of the second imaginary component of the $1-3-1$ HMLP (-) and the corresponding target versus the number
of testing patterns*

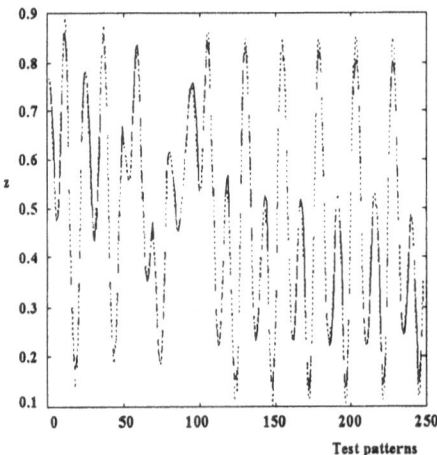

Figure 6.12: *Comparison between the trends of the third imaginary component
of the $1-3-1$ HMLP (-) and the corresponding target versus the number of
testing patterns*

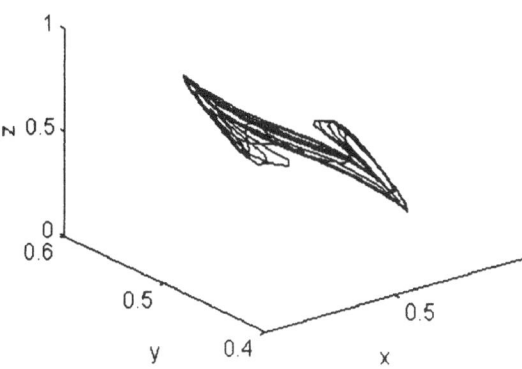

Figure 6.13: *Simulation of the attractor for Chua's circuit*

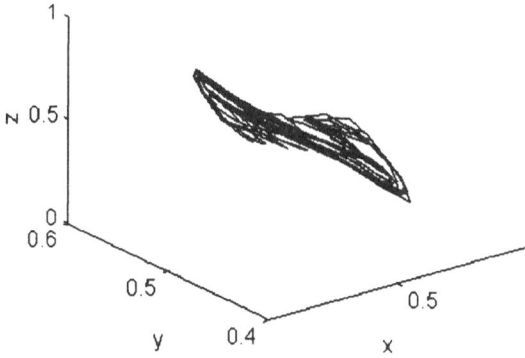

Figure 6.14: *Five-step-ahead prediction of Chua's attractor using a $1 - 3 - 1$ HMLP on 400 testing patterns*

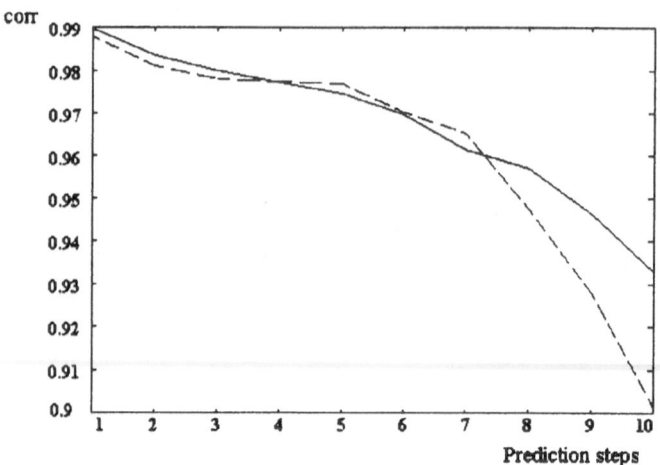

Figure 6.15: *Comparison between the correlation trends of the first output of the $4 - 8 - 4$ MLP (- -) and the real component of the $1 - 2 - 1$ HMLP (-) versus the prediction steps*

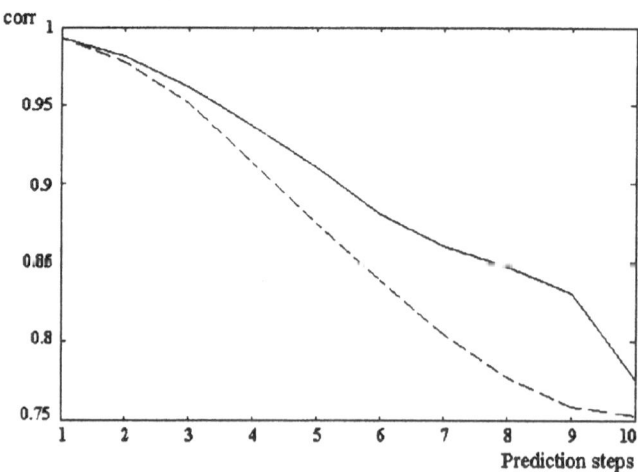

Figure 6.16: *Comparison between the correlation trends of the second output of the $4 - 8 - 4$ MLP (- -) and the first imaginary component of the $1 - 2 - 1$ HMLP (-) versus the prediction steps*

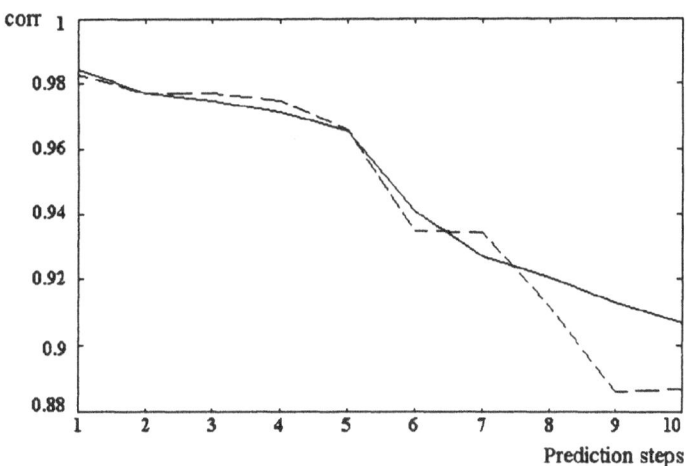

Figure 6.17: *Comparison between the correlation trends of the third output of the $4-8-4$ MLP (- -) and the second imaginary component of the $1-2-1$ HMLP (-) versus the prediction steps*

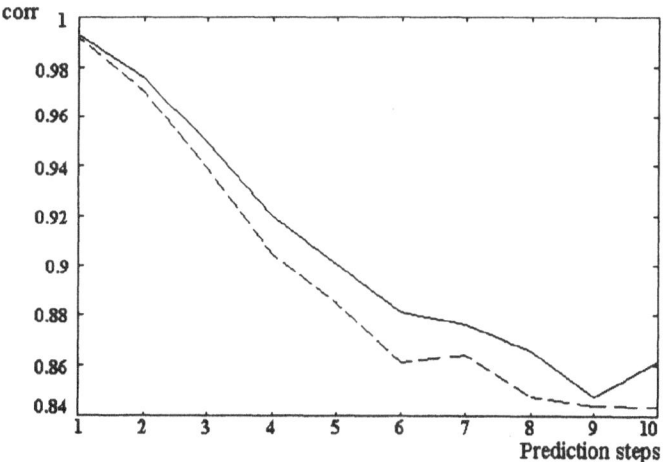

Figure 6.18: *Comparison between the correlation trends of the fourth output of the $4-8-4$ MLP (- -) and the third imaginary component of the $1-2-1$ HMLP (-) versus the prediction steps*

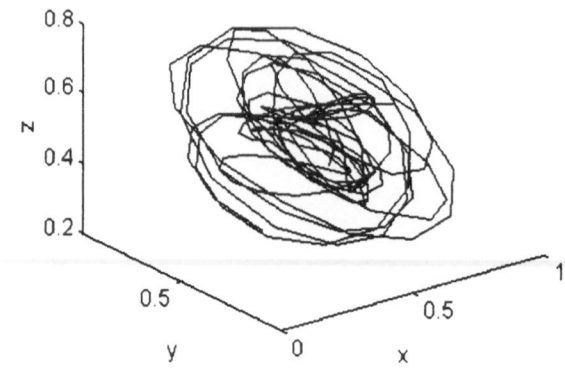

Figure 6.19: *Simulation of the attractor for Saito's circuit*

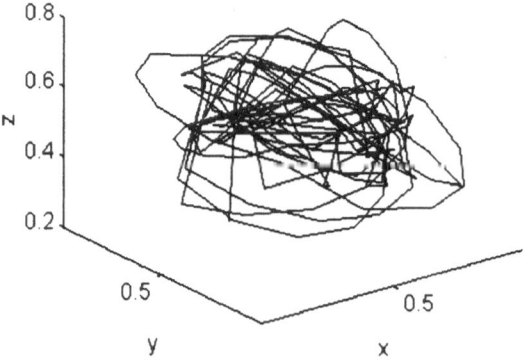

Figure 6.20: *Five step ahead prediction of Saito's attractor by using a $1-3-1$ HMLP on 400 testing patterns*

6.4.3   The Lorenz system

The set of equations known as the Lorenz system, [47] are reported as follows:

$$\dot{x} = \sigma(y - x) \tag{6.12a}$$

$$\dot{y} = -xz + rx - y \tag{6.12b}$$

$$\dot{z} = xy - bz \tag{6.12c}$$

The nominal values of the parameters are $\sigma = 10$, $r = \frac{8}{3}$ and $b = 28$. In the study, a set of real and hypercomplex topologies were trained with 200 terms of the time series. The best results were obtained with the real 3-11-3 MLP, (comparable with the 3-12-3 real MLP) and the 1-3-1 HMLP. A comparison between their performance, in terms of correlation the index for the state variables is shown in Fig.6.22-6.24 where the improvement due to the use of the HMLP can be appreciated.

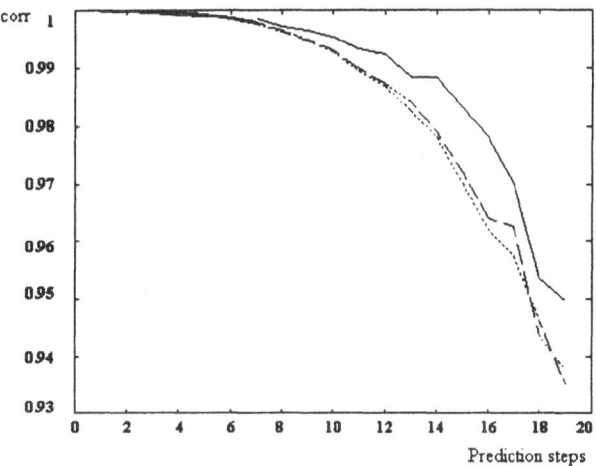

Figure 6.21: *Comparison between the trends of the first imaginary component of the $1-3-1$ HMLP (-) and the $3-11-3$ MLP (–) and the $3-12-3$ MLP (- ·)*

In Fig.6.25 and Fig.6.26 the simulated and the five-step-ahead predicted attractors of Lorenz system, computed on 500 testing patterns, are shown.

Once again the number of parameters employed by the HMLP is lower than that required by real MLP and the performance obtained is satisfactory.

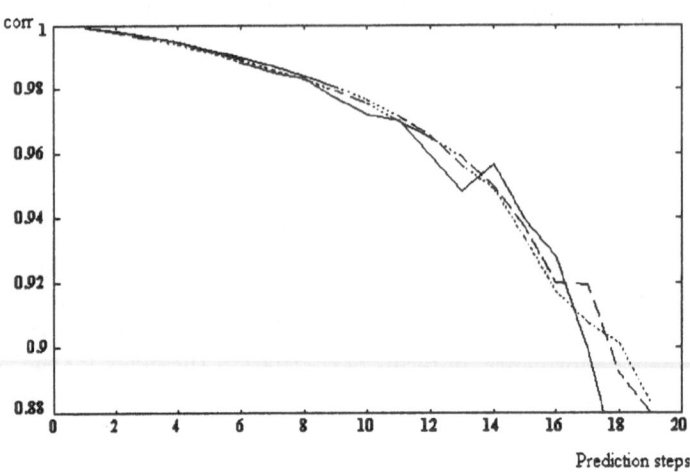

Figure 6.22: *Comparison between the trends of the second imaginary compo-nent of the $1-3-1$ HMLP (-) and the $3-11-3$ MLP (–) and the $3-12-3$ MLP (- ·)*

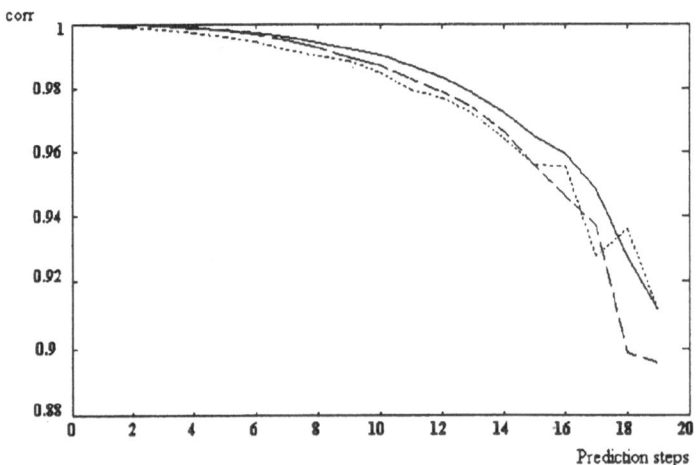

Figure 6.23: *Comparison between the trends of the third imaginary component of the $1-3-1$ HMLP (-) and the $3-11-3$ MLP (–) and the $3-12-3$ MLP (- ·)*

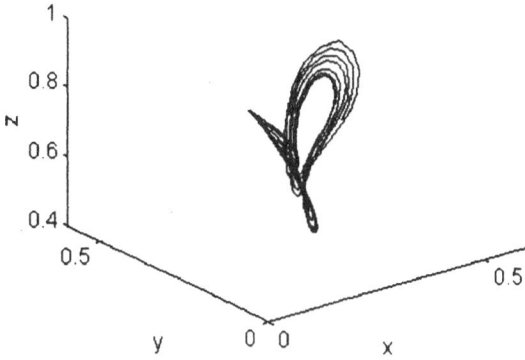

Figure 6.24: *Simulation of the attractor for the Lorenz circuit*

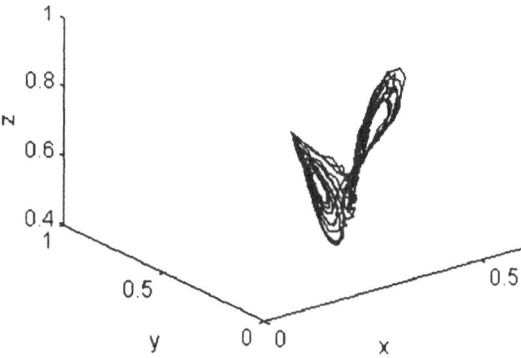

Figure 6.25: *Five-step-ahead prediction of Lorenz attractor using a* $1 - 3 - 1$ *HMLP on* 500 *testing patterns*

6.5   Conclusions

In this section the problem of chaotic time series prediction has been approached by using the multi-dimensional neural networks paradigms. Various examples are given to remark the suitability of the proposed approach. The results obtained are discussed emphasizing the improvements introduced by the hypercomplex neural networks. More specifically they consist in obtaining simpler structure, with few neurons and connections than the classical MLP architecture. From a numerical point of view, both for CMLPs and HMLPs the learning time (more specifically the number of learning cycles) is reduced owing to the lower dimension of the error surface (i.e. the lower number of connections), in particular the probability of getting stucked in local minima is decreased. A further advantage is given by the reduction of the number of products needed in the feed-forward phase when comparing a real and an hypercomplex networks with the same performance. The short-term prediction of canonical systems with chaotic behavior has been chosen as a benchmark to stress the suitability of the hypercomplex networks in solving difficult tasks. Both CMLP and HMLP can be successfully used for state estimation of non-linear (not necessarily chaotic) systems of great dimensions.

Chapter 7

Applications of quaternions in robotics

7.1 Introduction

In this chapter an application illustrating the growing interest in quaternion algebra in robotics is shown. This interest is mainly due to the simple representation of the rotations in 3-D space which can be obtained by using quaternion algebra and which leads to a reduced computational effort, to a compact and non-singular representation of the dynamic equations and to some advantages in proving the related stability theorems. The application presented in this chapter refers to the attitude control of a rigid body in a three-dimensional space, which consists of the computation of the control torque that allows to drive the object to a desired orientation and angular speed. Attitude control applications can be found in the control of satellites and robotic spacecrafts. Moreover, the orientation control of a rigid object held by robots can be translated into an attitude control problem [53], [51]. The classical approaches to the problem are feedback linearization or the Lyapunov-function based feedback controller. Due to the non-linearity of the model, they are not easy to apply. Efforts have concentrated on the choice of a simpler representation of the body dynamics and a great advantage can be introduced by using quaternion algebra [28], [33]. In the next section the quaternionic dynamic equations of a moving 3-D object are given. The following sections describe the neural approach to the attitude control problem and a second approach based on a PD controller coupled with a neural network. In both sections a numerical example is given.

7.2   Dynamic equations of a rigid body in quaternion algebra

The use of quaternion algebra in robotics allows an efficient representation of rotation, which introduces some advantages with respect to classical rotation matrices or Euler angles. In particular it can be shown that, by introducing the quaternionic representation of rotations, a mechanical system involving arbitrary rotations between 3-D reference frames does not degenerate for any configuration. Moreover, the computational complexity is greatly reduced, both as regards the number of parameters employed to represent a rotation and the number of sums and products necessary to compute the fundamental kinematic and dynamic quantities [54], [33].

7.2.1   Quaternionic representation of rotations

Let us consider the set of Euler parameters:

$$\mathbf{p} = [e_0, e_1, e_2, e_3]^T = [cos(\frac{\theta}{2}), sin(\frac{\theta}{2})u_x, sin(\frac{\theta}{2})u_y, sin(\frac{\theta}{2})u_z]^T \qquad (7.1)$$

representing the rotation of an angle θ around an axis with directors cosine u_x, u_y, u_z. The following relation holds:

$$e_0^2 + e_1^2 + e_2^2 + e_3^2 = 1$$

Euler parameters are a unit quaternion, in fact:

$$\mathbf{p} = cos(\frac{\theta}{2}) + sin(\frac{\theta}{2})\mathbf{u} \qquad (7.2)$$

where:

$$e_0 = cos(\frac{\theta}{2}), \quad sin(\frac{\theta}{2})^2 = \mathbf{e}^T\mathbf{e} \quad \text{and} \quad \mathbf{u} = \frac{\mathbf{e}}{sin(\frac{\theta}{2})}$$

If α is a quaternion, the operation:

$$\alpha' = \mathbf{p} \otimes \alpha \otimes \mathbf{p}^* \qquad (7.3)$$

transforms α into α' without modifying its scalar part. The vector part, on the contrary, is rotated by θ around the axis $\mathbf{u}$. In particular, if α is a quaternion vector, the above formula represents the statement of Euler's theorem which asserts that a generic rotation of a vector in 3-D space can be expressed by a single rotation of an angle θ around an axis $\mathbf{u}$. The inverse rotation is simply derived from:

$$\alpha = \mathbf{p}^* \otimes \alpha' \otimes \mathbf{p} \qquad (7.4)$$

Moreover:

$$\alpha' = \mathbf{p} \otimes \alpha \otimes \mathbf{p}^* = p^+\alpha \otimes p^* = p^{-*}p^+\alpha = p^{-T}p^+\alpha = A\alpha$$

where:

$$p^+ = \begin{bmatrix} p_0 & -\mathbf{p}^T \\ \mathbf{p} & p_0 I + \tilde{p} \end{bmatrix}$$

and

$$p^- = \begin{bmatrix} p_0 & -\mathbf{p}^T \\ \mathbf{p} & p_0 I - \tilde{p} \end{bmatrix}$$

with:

$$\tilde{p} = \begin{bmatrix} 0 & -p_3 & p_2 \\ p_3 & 0 & -p_1 \\ -p_2 & p_1 & 0 \end{bmatrix}$$

The matrix $A = p^{-T} p^+$ is therefore a $4x4$ rotational transformation matrix. In order to derive the connection between a generic rotation matrix and the corresponding quaternion, let us consider the matrix representing the rotation of an angle θ of a moving reference frame around an axis k given as:

$$R(k, \theta) = \begin{bmatrix} k_x^2 v\theta + c\theta & k_x k_y v\theta - k_z s\theta & k_x k_z v\theta + k_y s\theta \\ k_x k_y v\theta + k_z s\theta & k_y^2 v\theta + c\theta & k_y k_z v\theta - k_x s\theta \\ k_x k_z v\theta - k_y s\theta & k_y k_z v\theta + s\theta & k_z^2 v\theta + c\theta \end{bmatrix}$$

$$= \begin{bmatrix} n_x & o_x & a_x \\ n_y & o_y & a_y \\ n_z & o_z & a_z \end{bmatrix}$$

where k_x, k_y, k_z are the unit vector of the rotation axis, $v\theta = 1 - cos\theta$, $c\theta = cos\theta$ and $s\theta = sen\theta$. Equating the coefficients and solving the corresponding equations system the values of k_x, k_y, k_z and θ, and therefore the correspnding quaternion, are obtained. The following formulas are derived [54]:

$$k_x = \frac{o_x - a_y}{2sen\theta} \tag{7.5}$$

$$k_y = \frac{a_x - n_z}{2sen\theta} \tag{7.6}$$

$$k_z = \frac{n_y - o_z}{2sen\theta} \tag{7.7}$$

$$\theta = arctg \frac{\sqrt{((o_x - a_y)^2 + (a_x - n_z)^2 + (n_y - o_z)^2)}}{trace(R) - 1} \tag{7.8}$$

The inverse transformation is computed as:

$$R = \begin{bmatrix} q_0^2 + q_1^2 - q_2^2 - q_3^2 & 2(q_1 q_2 + q_0 q_3) & 2(q_1 q_3 - q_0 q_2) \\ 2(q_1 q_2 - q_0 q_3) & q_0^2 - q_1^2 + q_2^2 - q_3^2 & 2(q_2 q_3 + q_0 q_3) \\ 2(q_1 q_3 + q_0 q_2) & 2(q_2 q_3 - q_0 q_1) & q_0^2 - q_1^2 - q_2^2 + q_3^2 \end{bmatrix} \tag{7.9}$$

where q_0, q_1, q_2, q_3 are the components of the quaternion. A more compact representation is (Rodriquez formula):

$$R = (q_0^2 - \vec{q}^T \vec{q})I + 2(\vec{q}\vec{q}^T + q_0 \tilde{q}) \tag{7.10}$$

where I is the identity matrix.

7.2.2   Angular speed

Let us consider a rigid body moving in 3-D space and let $\mathbf{p}_{io}$ be the unit quaternion specifying the orientation of the body frame $I = (x_i, y_i, z_i)$ with respect to an inertial one $O = (x_o, y_o, z_o)$ and $\vec{r}$ a quaternion vector [33]. Let us denote with $\mathbf{r}^o$ the representation of $\vec{r}$ in the frame O and with $\mathbf{r}^i$ the representation of $\vec{r}$ in the frame I. Considering ((7.3)) we have:

$$\mathbf{r}^o = \mathbf{p}_{io} \otimes \mathbf{r}^i \otimes \mathbf{p}_{io}^* \tag{7.11}$$

$$\mathbf{r}^i = \mathbf{p}_{io}^* \otimes \mathbf{r}^o \otimes \mathbf{p}_{io} \tag{7.12}$$

Taking the derivative and making the products we get:

$$\dot{\mathbf{r}}^o = 2a \times \mathbf{r}^o + \mathbf{p}_{io} \otimes \dot{\mathbf{r}}^i \otimes \mathbf{p}_{io}^* = \tilde{\Omega}^o \mathbf{r}^o + A\dot{\mathbf{r}}^i \tag{7.13}$$

where: $\alpha = [0 \; a^T] = \dot{\mathbf{p}}_{io} \otimes \mathbf{p}_{io}^*$ and $\Omega^o = 2\alpha$ is the angular speed of the reference frame I with respect to O, represented in the frame O. It should be observed that the vector product $2a \times \mathbf{r}^o$ can be substituted, owing to the properties of quaternion algebra, by the matrix product $\tilde{\Omega}^o \mathbf{r}^o$, which can be inverted. Moreover:

$$\dot{\mathbf{r}}^i = -\tilde{\Omega}^i \mathbf{r}^o + A^T \dot{\mathbf{r}}^o = 2b \times \mathbf{r}^o + \mathbf{p}_{io}^* \otimes \dot{\mathbf{r}}^o \otimes \mathbf{p}_{io} = -\tilde{\Omega}^i \mathbf{r}^o + A^T \dot{\mathbf{r}}^o \tag{7.14}$$

where: $\beta = \dot{\mathbf{p}}_{io}^* \otimes \mathbf{p}_{io} = [b_o \; b^T]$ e $\Omega^i = -2\beta$ is the angular speed of I with respect to O, expressed in the coordinates I.
The following relations also hold:

$$\tilde{\Omega}^o = A\tilde{\Omega}^i A^T$$

$$\Omega^o = A\Omega^i$$

7.2.3   Angular acceleration

The angular acceleration of frame I with respect to O, represented in the co-ordinates O, is computed by differentiating the following relation:

$$\Omega^o = 2\dot{\mathbf{p}}_{io} \otimes \mathbf{p}_{io}^* \tag{7.15}$$

The results is:

$$\dot{\Omega}^o = 2(\ddot{\mathbf{p}}_{io} \otimes \mathbf{p}_{io}^* + \dot{\mathbf{p}}_{io} \otimes \dot{\mathbf{p}}_{io}^*) \tag{7.16}$$

And:

$$\dot{\Omega}^i = -2(\ddot{\mathbf{p}}_{io}^* \otimes \mathbf{p}_{io} + \dot{\mathbf{p}}_{io}^* \otimes \dot{\mathbf{p}}_{io}) \qquad (7.17)$$

Moreover:

$$\dot{\Omega}^o = A\dot{\Omega}^i \qquad (7.18)$$

Both the angular speed and the acceleration are vector quaternions.

7.2.4 Angular moment

The inertia matrix of a body of mass m with respect to the frame $O = (x_o, y_o, z_o)$, is defined as:

$$I^o = \int (r^T.rI_3 - rr^T)dm$$

where r is the vector denoting the distance between the mass dm and O, I_3 is the 3 by 3 identity matrix. In quaternion algebra the same definition can be adopted, replacing r with $\mathbf{r} = [0 \; r]$ and I_3 with I_4. The following inertia matrix is obtained:

$$\hat{I}^o = \begin{bmatrix} k & 0^T \\ 0 & I^o \end{bmatrix} \qquad (7.19)$$

where:

$$k = \frac{1}{2}trace(I^o)$$

If a torque $\mathbf{T}_{io}^o$ is acting on the body we have:

$$\mathbf{T}_{io}^o = \dot{\mathbf{H}}_{io}^o \qquad (7.20)$$

where $\mathbf{H}_{io}^o$ is the angular moment with respect to the center of moments and is given by:

$$\mathbf{H}_{io}^o = \hat{I}^o \Omega_{io}^o = 2\hat{I}^o(\dot{\mathbf{p}}_{io} \otimes \mathbf{p}_{io}^*) \qquad (7.21)$$

therefore:

$$\mathbf{T}_{io}^o = \hat{I}^o \dot{\Omega}^o + \tilde{\Omega}^o \hat{I}^o \Omega^o = I^o \dot{\omega}^o + \omega^o \times I^o \omega^o \qquad (7.22)$$

where ω is the vector part of Ω.

7.3 Neural attitude control in quaternion algebra

In this section a new approach to the solution of the attitude control problem is presented. As previously outlined the problem consists of finding the torque values able to lead the rigid body to the desired orientation and angular speed starting from any initial condition, in a fixed number of time steps.

The equations representing the attitude of a rigid body subject to an external

torque in quaternion algebra, as derived in the previous section, can be also
written after some manipulations as: [55]:

$$\dot{\mathbf{p}} = \frac{1}{2}[-\omega^T \vec{p}, \ p_0\omega + \vec{p} \times \omega]^T \qquad (7.23)$$

$$I\dot{\omega} = \tau - \omega \times I\omega \qquad (7.24)$$

where $\mathbf{p}$ is the quaternion representing the attitude of the body, ω and τ are
the vector quaternions representing the angular speed and the external torque
respectively and I is the matrix of principal moments of inertia of the body.
More specifically:

$$\mathbf{T}_{io}^o = [0 \quad \omega^T]$$

and

$$\Omega^o = [0 \quad \tau]^T$$

The approach to attitude control introduced in this chapter is based on a strategy proposed in [56] and suitably extended to quaternion algebra. Following
this approach in the first step a neural network, in this case an HMLP, is trained
to learn the state equations of the body. The HMLP obtained is introduced in
the more complex structure shown in Fig.7.1.

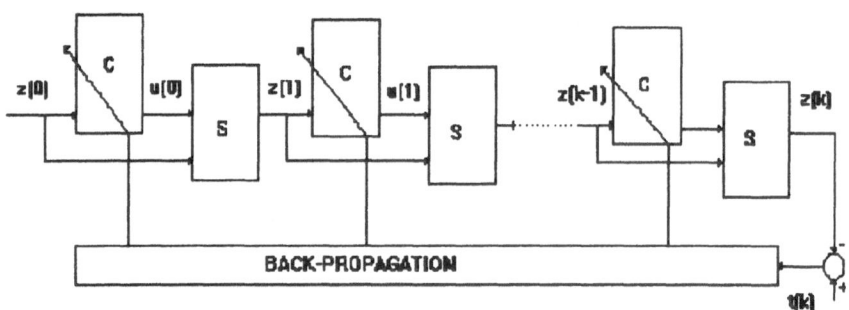

u=output of the controller
z=Inputs of the system
t=target
k=number of control steps

Figure 7.1: *MLP based structure for the attitude control of the rigid body*

In the figure the block 'S' is the previously trained HMLP emulating the
system and 'C' is another HMLP which represents the controller and has to
be trained, with a suitable learning algorithm, in order to obtain the desired
attitude in k time steps, where k is the number of C-S couples. The network
of Fig.7.1 is therefore trained maintaining the weights of the blocks S fixed

while the weights of the networks C are computed in order to minimize, with a gradient descent technique, a suitable cost function like the quadratic error between the output of the network and the desired target, on the whole set of patterns. The set of patterns used during the learning phase is made up of pairs of initial state-desired state values for the system to be controlled, where the state is represented by both the quaternionic orientation and the angular speed. During the training phase all the blocks C are maintained identical i.e. their weights are updated simultaneously. After the training phase, the controller is determined and acts on the system as a classical feedback controller. A detailed description of this method in real algebra can be found in [56]. In this application the block C has two inputs, which are the quaternions representing the orientation and the angular speed of the body, and one output which represents the control torque from the system. The block S has three inputs which are the orientation, the angular speed and the applied torque at time t ($t = 1 \ldots k$) and two outputs, representing the orientation and the angular speed at time $t + 1$. The learning algorithm derived for the proposed structure in quaternion algebra is given in [57]. For the sake of simplicity, only networks with one hidden layer and linear output neurons have been considered. The algorithm is based on the minimization, with a gradient descent technique, of the error:

$$E = \sum_{i=1}^{nos}(\mathbf{t}_i - \mathbf{y}_i)^2 \qquad (7.25)$$

where $\mathbf{t}_i$ and $\mathbf{y}_i$ are the desired and actual values respectively of the i-th output units of the network representing the system and *nos* is the number of output of the system.

To underline the suitability of the proposed approach a spacecraft example taken from [55] is considered.

The inertia matrix considered is $I = diag(1, 0.83, 0.92)\ kg\ m^2$. In a first phase the HMLP which simulates the rigid body dynamics (the block S in Fig.7.1) was trained to learn the behaviour of a discretized version of the system, using a sampling time of $0.01sec$. The set of learning and testing patterns was generated starting from a set of random initial orientation, angular speeds and input torques. 150 patterns were used during the learning phase, while 350 were used to test the generalization capability of the network. The best results were obtained using an HMLP with one hidden layer consisting of 3 neurons. In Fig.7.2 a comparison is made between the output of the HMLP and the target for one component of the output while in Fig.7.3 the corresponding error for the whole set of testing patterns is shown.

In the second phase the HMLP implementing the controller has to be trained to drive the system from any initial condition to the zero condition (orientation of the body reference aligned with the fixed frame and zero angular speed) in a fixed number of time steps. It should be observed that the range of initial conditions has to be chosen according to the maximum allowable torque and the desired number of time steps. For the example considered 5 time steps and

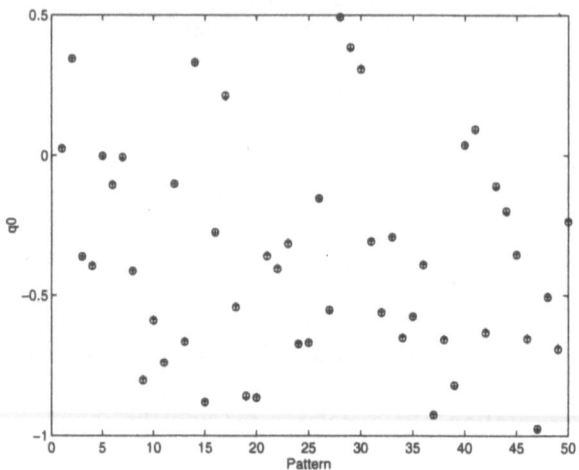

Figure 7.2: *Neural controller synthesis: comparison between the output (o) of the HMLP and its target(+) for the first component of the angular orientation*

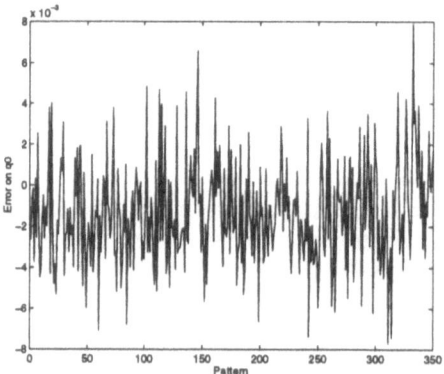

Figure 7.3: *Neural controller synthesis: error between the output of the HMLP and its target for the first component of the angular orientation*

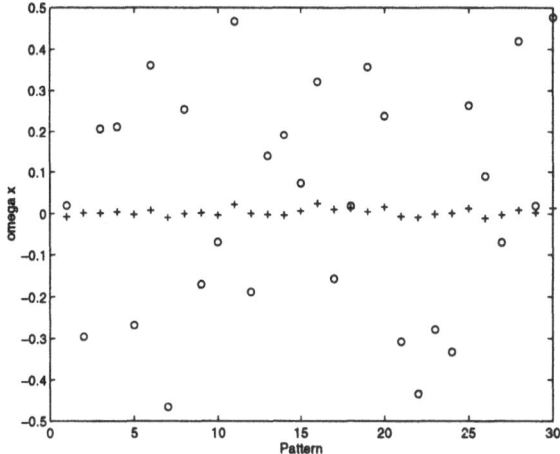

Figure 7.4: *Neural controller synthesis: initial (o) and final (+) values of the first component of the angular speed obtained with the neural controller for 30 testing patterns*

an HMLP controller with 2 hidden neurons were adopted. A set of 50 initial conditions were used for the learning phase while 150 were used to test the controller performances. Fig.7.4 shows the initial and final angular speed values (first component) for 30 testing patterns. It can be observed that the controller is able to drive the system to the zero condition even with initial conditions not considered during the learning phase, owing to the generalization capabilities of the HMLP.

To underline the advantages introduced by the HMLP with respect to the corresponding real structure it should be considered that each quaternionic unit corresponds to four real units. The total number of real parameters (each quaternion is evalued as 4 real parameters), involved in the structure with 3 hidden neurons for the HMLP of block S and 2 hidden neurons in the C blocks, is 116. The corresponding structure, which employs the real MLP, has 10 inputs and 7 outputs for the system emulator and 7 inputs and 3 outputs for the controller. The corresponding numbers of hidden units are 12 and 8 respectively. Each C-S block therefore includes 314 parameters against the 116 of the HMLP-based structure. It is therefore clear that the training phase of the control structure in quaternion algebra is much more unexpansive than the corresponding neural architecture in real algebra. As can be observed the considered strategy gives good results. However, when the number of time steps increases, the computational complexity can be overwhelming. To overcome this problem a different approach is proposed in the next section.

7.4   A quaternionic PD controller with a feedforward neural term

The proposed strategy is based on two concurrent controllers. The first one is a feedback-type PD controller based on the quaternionic representation of the dynamic equations, while the second is a feedforward controller implemented by using an HMLP. The feedback controller is computed according to the results reported in [53], where it was shown that a PD controller is globally stabilizing for a class of desired trajectories. Moreover the stability of such a PD controller is model-independent. The control law proposed in [53] is:

$$\tau_{PD} = k_p \vec{q} - k_v \Delta\omega \qquad (7.26)$$

where $\vec{q}$ and $\Delta\omega$ represent the error between the actual and desired attitude respectively, while k_p and k_v are positive scalar gains. In particular $\vec{q}$ is the vector quaternion denoting the orientation of the desired attitude with respect to the fixed body reference frame and is computed as:

$$\mathbf{q} = \mathbf{q_{do}} \otimes \mathbf{q_{so}^{-1}} \qquad (7.27)$$

where $\mathbf{q}_{do}$ and $\mathbf{q}_{so}$ are the quaternions representing the desired attitude and the body frame with respect to the inertial fixed frame, while the speed error is:

$$\omega = \omega_{so} - \omega_{do}. \qquad (7.28)$$

Here again, without loss of generality it is assumed that the desired attitude corresponds to the zero attitude, i.e. the body frame aligned with the inertial frame and zero angular speed. The class of desired trajectories is given by:

$$\theta_d(t) = \theta_i exp(-\alpha t^2) \qquad (7.29)$$

$$\omega_d(t) = -2\,\alpha\,\theta_i\,t\,exp(-\alpha t^2) \cdot \hat{k} \qquad (7.30)$$

where θ_i is the initial rotation of the body along the axis $\hat{k}$ and α is a positive gain. The results of [53] are extended introducing a feedforward term in the controller. The new control law becomes:

$$\tau = \tau_{PD} + \tau_{ff} = K_p \vec{q} - K_v \Delta\omega + \tau_{ff}(\mathbf{q}_{do}, \omega_{do}) \qquad (7.31)$$

where the feedforward term $\tau_{ff}(\mathbf{q}_{do}, \omega_{do})$ is implemented by using an HMLP neural network. The neural feedforward term allows better compensation of the nonlinearities of the model without compromising the closed loop stability. Moreover the network is adapted to the system without any a priori knowledge of the body parameters, allowing the controller performance to be improved even with initial conditions not considered in the learning phase. Following the approach proposed in [52], during the learning phase the feedback term

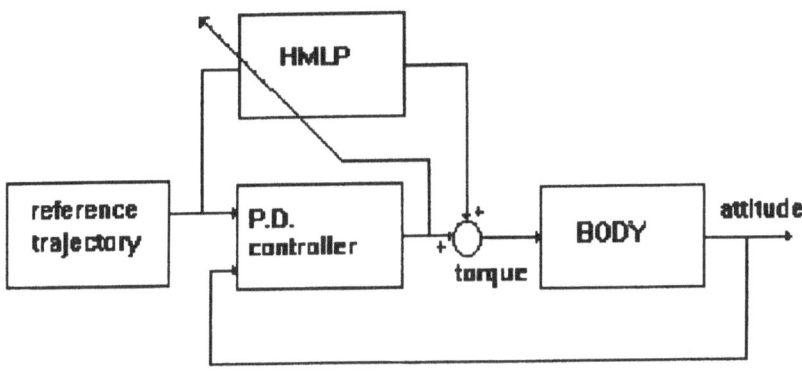

Figure 7.5: *Block scheme of the control structure*

τ_{PD} was adopted as the learning error of the network. A block diagram of the whole control scheme is shown in Fig.7.5. The feedforward term $\tau_{ff}(\mathbf{q}_{do}, \omega_{do})$ is realized by using an HMLP with two input neurons ($\mathbf{q}_{do}, \omega_{do}$) and one output neuron τ_{ff}. The HMLP was trained according to the following steps:

- The HMLP weights are randomly initialized.

- A set of N trajectories is generated, starting with different initial conditions according to equations (7.29) and (7.30).

- For each trajectory the control law (7.31) is applied to the system.

- During the control action, at fixed time steps, a learning cycle for the HMLP is performed using τ_{PD} as the error for the back-propagation algorithm.

- The last two steps are repeated until a good tracking error is obtained on the whole set of trajectories.

In order to show the suitability of the proposed approach the same system and PD controller reported in [53] were considered as an example.

In particular the inertia matrix adopted is $I = diag(1 , 0.63 , 0.85)$ and the PD controller gains are chosen as $k_p = 4$, $k_v = 8$. The control action has been applied for $T = 10$ sec for each trajectory and the learning cycles for the HMLP were computed with a time step $\Delta T = 0.1$ sec. As regards the HMLP topology, a network with one hidden layer and 2 hidden neurons performed well.

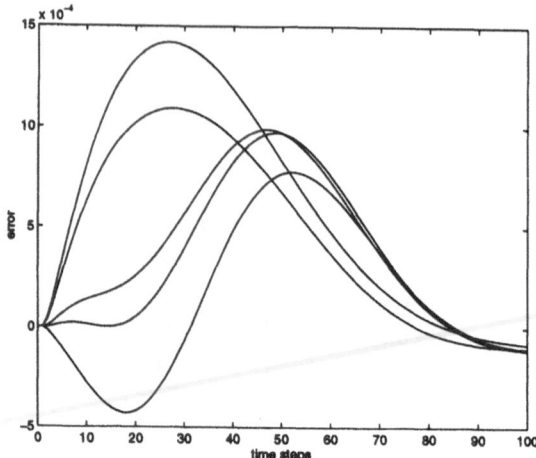

Figure 7.6: *Simulation of the PD and neural controller closed loop system: scalar part of the error between the actual and the desired trajectories for 5 trajectories not used during the learning phase*

In Fig.7.6 the scalar part of the quaternion representing the error between the actual and the desired trajectories is given for 5 trajectories not presented to the system during the learning phase. As can be observed from the figure the controlled system once more performs well. In Fig.7.7 a comparison is made for a single trajectory between the scalar parts of the error quaternions, when a single PD controller and the proposed PD+HMLP controller are connected to the system.

Again the improvement introduced by the neural feedforward term is remarkable. Similar results were obtained for the whole set of unknown trajectories.

7.5   Conclusions

In this section an innovative application of hypercomplex neural networks in robotics has been introduced. While the role of quaternions in robot control has been widely dealt with, this topic has not been investigated previously. Two approaches have been introduced and a critical comparison has been performed. The efficiency of the HMLP with respect to the classical MLP has been emphasized and the advantages in terms of simplicity and numerical efforts of the learning algorithm are also in agreement with the results discussed in chapter 6. The adopted control strategy is quite general and suitable for each problem of attitude control. Moreover the new opportunity for modelling the dynamics of rigid body by using quaternions based neural networks emerged from the reported discussion.

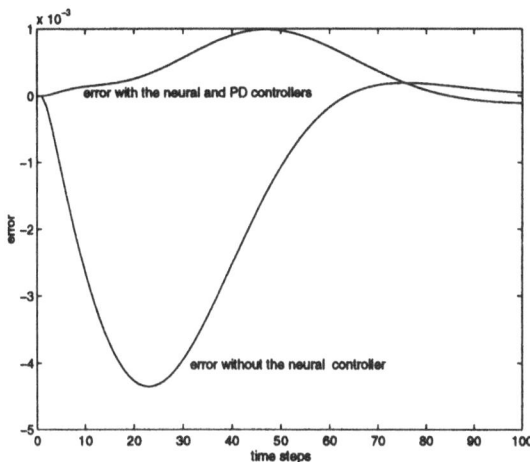

Figure 7.7: *Simulation of the PD and neural controller closed loop system: comparison between the scalar parts of the error on a single trajectory with and without the neural controller.*

Appendix A

Basic definitions

A.1 Definitions on algebras

Quaternion:
$$\mathbf{q} = q_0 + q_1\mathbf{i} + q_2\mathbf{j} + q_3\mathbf{k} = q_0 + \vec{q} \tag{A.1}$$

Quaternion components:
$$q_0, \quad q_1, \quad q_2, \quad q_3 \tag{A.2}$$

Imaginary units of a quaternion:
$$\mathbf{i}, \ \mathbf{j}, \ \mathbf{k} \tag{A.3}$$

Scalar quaternion:
$$q_0 \tag{A.4}$$

Vector quaternion:
$$\vec{q} \tag{A.5}$$

Conjugate quaternion:
$$q^* = q_o - \vec{q} = q_0 - q_1\mathbf{i} - q_2\mathbf{j} - q_3\mathbf{k} \tag{A.6}$$

Quaternion modulus:
$$|\mathbf{q}| = \sqrt{\mathbf{q}\mathbf{q}^*} \tag{A.7}$$

Matrix representation of a quaternion:
$$\mathbf{q} = [q_0, \ q_1, \ q_2, \ q_3]^T \tag{A.8}$$

Sum:
$$\mathbf{q} + \mathbf{p} = (q_0 + p_0) + (q_1 + p_1)\mathbf{i} + (q_2 + p_2)\mathbf{j} + (q_3 + p_3)\mathbf{k} \tag{A.9}$$

Subtraction:
$$\mathbf{q} - \mathbf{p} = (q_0 - p_0) + (q_1 - p_1)\mathbf{i} + (q_2 - p_2)\mathbf{j} + (q_3 - p_3)\mathbf{k} \tag{A.10}$$

Product:

$$\mathbf{q} \otimes \mathbf{p} = q_0 p_0 - \vec{p} \cdot \vec{p} + q_0 \vec{p} + p_0 \vec{q} + \vec{q} \times \vec{p} = \begin{bmatrix} q_0 p_0 - \vec{q}^T \vec{p} \\ \vec{q} p_0 + (q_0 I + \tilde{q}) \vec{p} \end{bmatrix} \qquad (A.11)$$

where $\tilde{q}$ is the matrix:

$$\tilde{q} = \begin{bmatrix} 0 & -q_3 & q_2 \\ q_3 & 0 & -q_1 \\ -q_2 & q_1 & 0 \end{bmatrix} \qquad (A.12)$$

Products among the imaginary units:

$$\mathbf{i}^2 = \mathbf{j}^2 = \mathbf{k}^2 = -1 \qquad (A.13)$$

$$\mathbf{jk} = -\mathbf{kj} = \mathbf{i} \qquad (A.14)$$

$$\mathbf{ki} = -\mathbf{ik} = \mathbf{j} \qquad (A.15)$$

$$\mathbf{ij} = -\mathbf{ji} = \mathbf{k} \qquad (A.16)$$

Inverse:

$$\mathbf{q}^{-1} = \frac{\mathbf{q}^*}{\mathbf{q}^* \otimes \mathbf{q}} \qquad (A.17)$$

A.2 Basic definitions on functions

Definition A.2.1 *Density in* $\Re$.
A subset S of $C(I_n)$ is dense *in $C(I_n)$ if $\forall f \in C(I_n)$ and $\forall \epsilon > 0$, $\exists G(x) \in S$ such that:*

$$\| G(\bar{x}) - f(\bar{x}) \| < \epsilon \ \ \forall \bar{x} \in I_n$$

Definition A.2.2 *Density in* C.
A subset S of $C(I_n)$ is dense *in $C(D_n)$ if $\forall f \in C(D_n)$ e $\forall \epsilon > 0$, $\exists G(x) \in S$ such that:*

$$\| G(z) - f(z) \| < \epsilon \ \ \forall z \in D_n$$

The norm used in the previous definitions is defined as follows:

Definition A.2.3 *Norm of a function.*
For a given function $f(x)$, $x \in K$, the norm of $f(x)$ is defined as follows:

$$\| f(x) \| = sup_{x/in K} |f(x)|$$

Appendix B

Chaotic systems

B.1 Definition of chaotic system

A formal definition of chaotic system is based on the concepts of density, transitivity and sensitivity on initial conditions, introduced in the following [60].

Definition B.1.1 *Let us consider a vector space X in which a metrix d has been defined. A subset $B \subseteq X$ is said* **dense** *in X if the set of the* **accumulation points** *of B is equal to X. A sequence $\{x_n\}_{n=0}^{\infty}$ of points in X is dense in X if for any point y of X there exist a subsequence $\{x_{\sigma_n}\}_{n=0}^{\infty}$ converging to y. An orbit $\{x_n\}_{n=0}^{\infty}$ of a dynamical system f defined into the space X is dense in X if the sequence of points $\{x_n\}_{n=0}^{\infty}$ is dense in X.*

Definition B.1.2 *A dynamical system $\dot{x} = f(x)$ is called* **transitive** *if, for any pair of open sets A, $B \subset X$, there exist a finite value of n for which it holds:*

$$A \cap f^{\circ n}(B) \neq \oslash$$

Definition B.1.3 *Let $\dot{x} = f(x)$ be a dynamical system defined in a state space X and let d be the metrix defined in X. A dynamical system is called* **sensitive to the initial conditions** *if there exists $\delta > 0$ such that, for any $x \in X$ and any neighbourhood B of x of radius $\epsilon > 0$, there exists $y \in B$ and $n \geq 0$ such that $d(f^{\circ n}(x), f^{\circ n}(y)) > \delta$.*

The operator $f^{\circ n}$ is a transformation $f^{\circ n} : X \to X$ with the following properties: $f^{\circ 0}(x) = x$; $f^{\circ 1}(x) = f(x)$; $f^{\circ n}(x) = f(f^{\circ(n-1)}(x))$.

Definition B.1.4 *A dynamical system $\dot{x} = f(x)$, defined into the state space $X \subseteq R^n$ is chaotic if:*

- *is transitive;*

- *is sensitive to the initial conditions;*

- *the periodic solutions of f are dense in the state space.*

The time evolution trend of chaotic variables does not show any periodicity and looks like a stochastic behaviour. Moreover the spectrum does not show distinct peaks, but has a continuous shape, like a white noise spectrum. The sentitivity to the initial conditions makes the chaotic system behaviour unpredictable; in fact any imprecision, even due to finite word lenght effects, leads to completely different orbits.

The transitivity property enables a chaotic system to be decomposed into non interacting subsystems. One regularity element in chaotic systems can be found in the density of periodic points. This contraddiction between the divergent behaviour of trajectories starting from initial conditions arbitrarily next to one another and the finite volume containing the attractor cannot be solved by euclidean geometry; a satisfactory answer can be provided referring to *fractal geometry* .

In the following are reported the models and the circuit realizations for the chaotic systems introduced in Chap.6, for which some predictive models have been performed by using neural networks [59].

B.2   The Lorenz system

The equation system describing the dynamic of the Lorenz system was developed in order to investigate atmospheric phenomena. The study was based on the resemblance to the convective motion of fluids in a closed volume in presence of a thermal gradient [47]. The Lorenz system is composed of three non linear differential equations in which: x represents the convective motion intensity;

y is the thermal difference between the ascending and descending component; z represents the vertical thermal profile distorsion. The equations are the following:

$$\dot{x} = \sigma(y - x) \tag{B.1a}$$

$$\dot{y} = -xz + rx - y \tag{B.1b}$$

$$\dot{z} = xy - bz \tag{B.1c}$$

The parameters σ, r e b are proportional to the Prandtl number, to the Rayleigh number and to the physical dimensions of the fluid envelope. Their nominal value is $\sigma = 10$, $r = \frac{8}{3}$ e $b = 28$. The circuit scheme of the Lorenz system is reported in Fig. B.1.   For the realization operational amplifiers μA741 and TL082, and analog multipliers AD633JN were used. The other component values are reported in the following table.

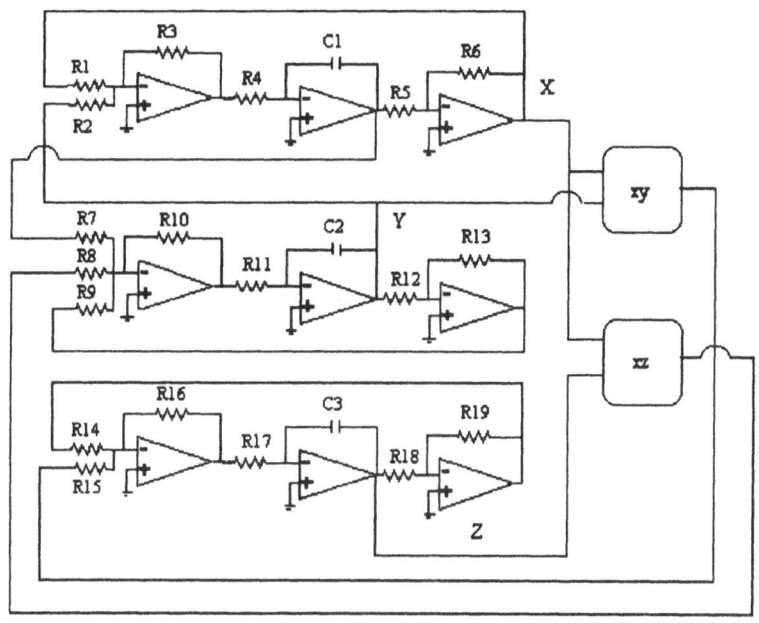

Figure B.1: *Circuit realization of the Lorenz system*

B.3 Chua's circuit

It is well known in literature because it represents the simplest circuit able to show chaotic motions [48]. It is described by the following non linear differential equations:

$$\dot{x} = \alpha(y - h(x)) \tag{B.2a}$$

$$\dot{y} = x - y + z \tag{B.2b}$$

$$\dot{z} = \beta y - \gamma z \tag{B.2c}$$

where:

$$h(x) = m_1 x + 0.5(m_0 - m_1)(\mid x + 1 \mid - \mid x - 1 \mid)$$

Lorenz circuit: component values
Resistors
R1=R2=R3=R4=R5=R6=R8=R10=R11=R12=R16=R17=10 $K\Omega$
R7=3.3 $K\Omega$
R9=R13=R14=R15=R19=1 $K\Omega$
R18=39 $K\Omega$
Capacitors
C1=C2=C3=100 nF

The well known 'double-scroll' attractor is obtained for the following parameter values:

$$(\alpha, \beta, \gamma, m_0, m_1) = (9, 14.286, 0, -1/7, 2/7)$$

In the circuit the non linear part is represented by the Chua's diode N_R. The circuit scheme of N_R and its characteristic are reported in Fig. B.2.   The

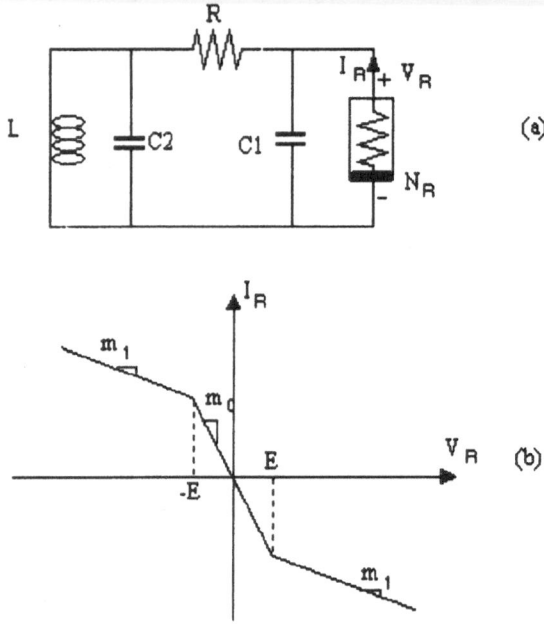

Figure B.2: *Chua's circuit (a) and Chua's diode characteristic (b)*

circuit realization of the non linear component was very easily realized by exploiting the natural saturations of two operational amplifiers with different gains, connected in parallel as shown in Fig. B.3.   The inductor $L = 18\ mH$ was realised by using a NIC (Negative Impedence Converter), whose scheme is shown in Fig. B.4; such a realization allows a sensitive reduction of the circuit dimensions.   The component values are reported in the following table.

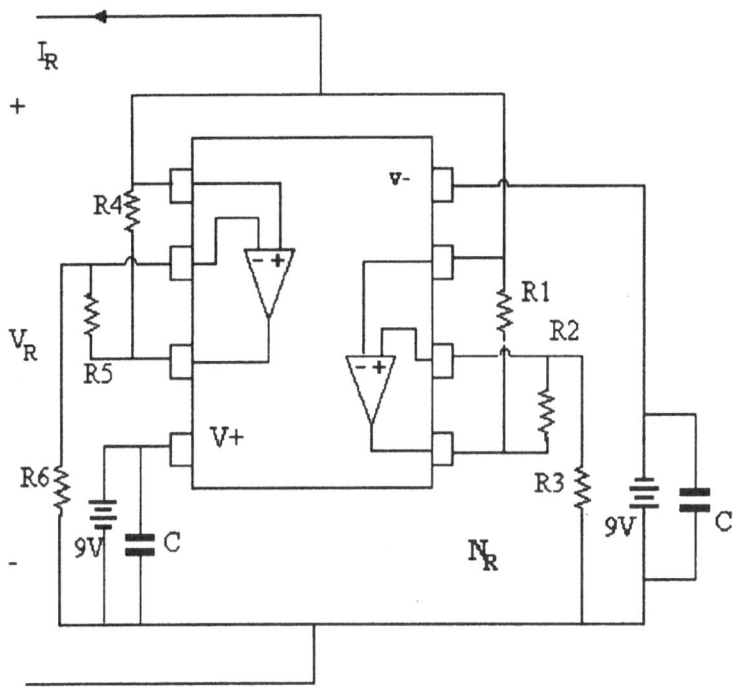

Figure B.3: *Circuit scheme of Chua's diode*

B.4   Saito's circuit

The chaotic circuit introduced in [49] is characterized by four state variables and by five parameters, whose variations allow the system to transit through several steady state behaviours, which include toroidal attractors and also more complex ones. Such a dynamics is due to the presence of a hysteresys controlled voltage source. The Saito circuit dynamics is described by the following equtions:

$$\begin{bmatrix} \dot{x}_1 \\ \dot{y}_1 \end{bmatrix} = \begin{bmatrix} -1 & 1 \\ -\alpha_1 & \alpha_1\beta_1 \end{bmatrix} \begin{bmatrix} x_1 - \eta p_1 h(z) \\ y_1 - \eta \frac{p_1}{\beta_1} h(z) \end{bmatrix} \qquad (B.3)$$

$$\begin{bmatrix} \dot{x}_2 \\ \dot{y}_2 \end{bmatrix} = \begin{bmatrix} -1 & 1 \\ -\alpha_2 & \alpha_2\beta_2 \end{bmatrix} \begin{bmatrix} x_2 - \eta p_2 h(z) \\ y_2 - \eta \frac{p_2}{\beta_2} h(z) \end{bmatrix} \qquad (B.4)$$

where:

$$h(z) = \begin{cases} 1 & \text{per } z \geq -1 \\ -1 & \text{per } z \leq 1 \end{cases}$$

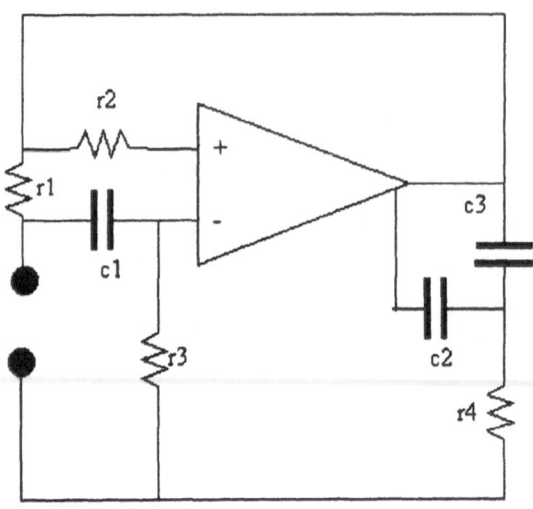

Figure B.4: *Circuit schema of a NIC*

is the hysteresis, and:

$$z = x_1 + x_2, \quad p_1 = \frac{\beta_1}{1 - \beta_1}, \quad p_2 = \frac{\beta_2}{1 - \beta_2}$$

The parameters α_1 and α_2 modulate the oscillation frequency, β_1 and β_2 control the damping factor, η is the hysteresis threshold amplitude. Referring to the parameters of the circuit shown in Fig.B.5, it also results:

$$x_1 = \frac{\eta v_1}{E} \quad x_2 = \frac{\eta v_2}{E}$$

$$y_1 = \frac{\eta R i_1}{E} \quad y_2 = \frac{\eta R i_2}{E}$$

Chua's circuit: component values		
Resistors		
R1=R2=220 Ω	R3=2.2 $K\Omega$	
R4=R5=22 $K\Omega$	R6=3.3 $K\Omega$	
r1 = 18 Ω, r2=r3=100 Ω	r4 = 10 Ω	
Tunable Resistor		
R=2 $K\Omega$		
Capacitors		
C1=10 nF	C2=100 nF	
c1 = 0.1 μF	c2=33 pF , c3=330 pF	

$$\alpha_1 = \frac{R^2 C}{L_1} \quad \alpha_2 = \frac{R^2 C}{L_2} \quad \beta_1 = \frac{r_1}{R} \quad \beta_2 = \frac{r_2}{R}$$

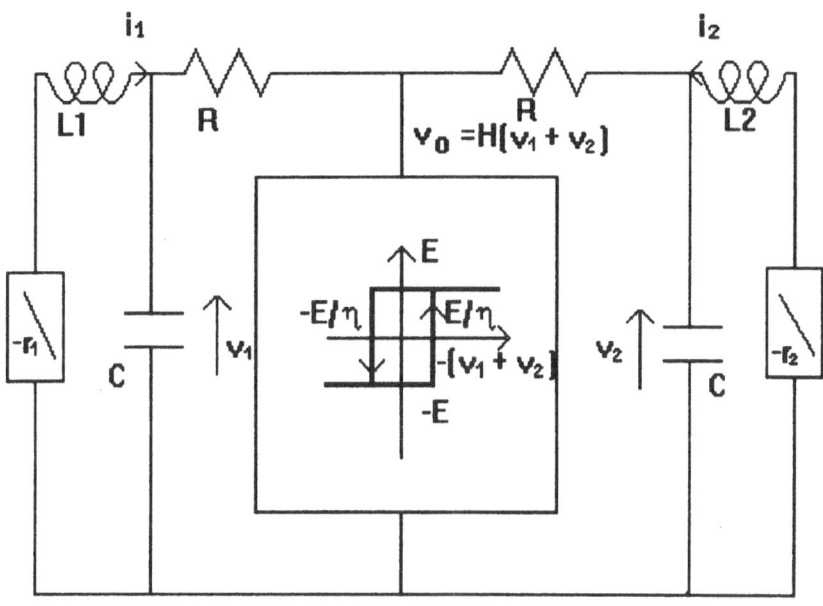

Figure B.5: *Saito chaotic circuit*

Appendix C

Getting started with HMLP Software

C.1 Introduction

This appendix is organized as follows. The first part contains the Matlab routines that implement the basic quaternionic operations and the relations between the rotation matrix and the quaternionic representation of rotations. In particular the following operations are contained:

- Quaternion product (prodquat.m);

- conjugate of a quaternion (quacon.m);

- inverse of a quaternion (qinv.m);

- norm of a quaternion (normq.m);

- quaternion corresponding to rotations of a given angle around a given axis (ax2q.m);

- rotations axis and angle corresponding to a quaternion (q2ax.m);

- quaternion corresponding to a given rotation matrix (m2q.m);

- rotation matrix corresponding to a given quaternion (q2m.m);

The second part contains the MATLAB routines implementing the Back Propagation algorithm for the HMLP and an example of function approximation by using HMLP. A block scheme represents all the subroutines used by the main program *hmlpmenu.m.*

The sequence of action to train an HMLP and test its performance are represented in a flow chart.

Moreover an example of training of an HMLP to approximate a simple quaternionic function with the program *hmlpmenu* is described in detail.

All the subroutines are reported at the end of the appendix.

C.2 Fundamental quaternionic operations

prodquat.m

```
function [q]=prodquat(q1,q2);
%PRODQUAT Quaternionic product:
% Q = Q1 (X) Q2
% PRODQUAT(Q1,Q2) is the product between two
% quaternions, Q1 and Q2.
%
% See also NORMQ, PRODVETT, QINV, QUACON,
% QUAPRODC.

q=[q1(1)*q2(1)-q1(2:4)'*q2(2:4);
q1(1)*q2(2:4)+q2(1)*q1(2:4)+prodvett(q1(2:4),q2(2:4))];
```

quacon.m

```
function [q]=quacon(q1);
%QUACON Conjugate of a quaternion.
%QUACON(Q1) returns the conjugate of the
%quaternion Q1.
%
%See also NORMQ, PRODQUAT, PRODVETT, QINV,
% QUAPRODC.

q=[q1(1); -q1(2:4,1)];
```

qinv.m

```
function qi=qinv(q);
%QINV Quaternion inverse.
%QINV(Q) returns the inverse of the quaternion Q.
%
%See also NORMQ, NORMVETT, PRODVETT, PRODQUAT,
% QUACON,QUAPRODC.


% Quadratic norm of the quaternion q.
a=normq(q);
% Quaternion inverse.
qi=quacon(q)./a(1);
```

normq.m

```
function [e]=normq(q);
%NORMQ Quaternion quadratic norm.
%NORMQ(Q) is the quadratic norm of the quaternion Q.
%The expression used to compute this norm is:
%              normq(Q) = Q (X) Q*
%where '*' denotes the conjugate operator.
%
%See also NORMVETT, PRODQUAT, PRODVETT, QINV,
% QUACON,QUAPRODC.
e=prodquat(q,quacon(q));
```

ax2q.m

```
function q=ax2q(ax, teta);
%AX2Q Representation of rotations using the
%Euler parameters, that are unit quaternions.
%
%AX2Q(AX,TETA) returns the quaternion corresponding
%to the rotation of an angle TETA around an axis AX.
%
%See also Q2AX, Q2AXV.

% Scalar quaternion of q obtained from the rotation
% angle "teta".
q(1,1)=cos(teta/2);

% Vector quaternion of q obtained from the rotation
%angle "teta" and from the rotation axis "ax".
q(2,1)=ax(1)*sin(teta/2);
q(3,1)=ax(2)*sin(teta/2);
q(4,1)=ax(3)*sin(teta/2);
```

m2q.m

```
function q=m2q(M);
%M2Q Quaternion corresponding to a rotation matrix.
%M2Q(M) is the quaternion obtained from the rotation
%matrix M.
%
%See also Q2M.

% Terms used in the four quadrant arctangent to compute
```

```
% the rotation angle "teta".
x=sqrt((M(3,2)-M(2,3))^2+(M(1,3)-M(3,1))^2+(M(2,1)-M(1,2))^2);
y=M(1,1)+M(2,2)+M(3,3)-1;

% Rotation angle of a moving reference frame.
teta=atan2(x,y);

% Components of the rotation axis of a moving
%reference frame.
ax(1,1)=(M(3,2)-M(2,3))/2*sin(teta);
ax(2,1)=(M(1,3)-M(3,1))/2*sin(teta);
ax(3,1)=(M(2,1)-M(1,2))/2*sin(teta);

%Quaternion corresponding to the rotation angle
%"teta" and to the rotation
%axis "ax"
q=ax2q(ax,teta);


q2ax.m

function [ax, teta]=q2ax(q);
%Q2AX Rotation axis and angle corresponding to a quaternion.
%The syntax of the function Q2AX is:
%[AX, TETA]=Q2AX(Q)
%AX is the rotation axis and TETA is the rotation angle
%corresponding to the quaternion Q.
%
% See also AX2Q, Q2AXV.


% A quaternion can be expressed as follows:
%    q = cos(teta/2) + sin(teta/2)*ax

% Vector quaternion of q.
e=[q(2) q(3) q(4)]';
% Square of sin(teta/2).
x=sqrt(e'*e);
% Scalar quaternion of q.
y=q(1);

% Rotation angle of a reference frame.
teta=2*atan2(x,y);
% Rotation axis of a reference frame.
ax=e/x;
```

q2m.m

```
function M=q2m(q);
%Q2M Rotation matrix corresponding to a quaternion.
%Q2M(Q) returns a rotation matrix that is obtained from the
%quaternion Q.
%
% See also M2Q.

M=zeros(3,3);
% Elements of the rotation matrix obtained from the
%quaternion q.
M(1,1)=q(1)*q(1)+q(2)*q(2)-q(3)*q(3)-q(4)*q(4);
M(1,2)=2*(q(2)*q(3)+q(1)*q(4));
M(1,3)=2*(q(2)*q(4)-q(1)*q(3));
M(2,1)=2*(q(2)*q(3)-q(1)*q(4));
M(2,2)=q(1)*q(1)-q(2)*q(2)+q(3)*q(3)-q(4)*q(4);
M(2,3)=2*(q(3)*q(4)+q(1)*q(2));
M(3,1)=2*(q(2)*q(4)+q(1)*q(3));
M(3,2)=2*(q(3)*q(4)-q(1)*q(2));
M(3,3)=q(1)*q(1)-q(2)*q(2)-q(3)*q(3)+q(4)*q(4);
```

C.3   Flow Charts

In the following, block schemes and flow charts of the HMLP learning algorithm
are reported.

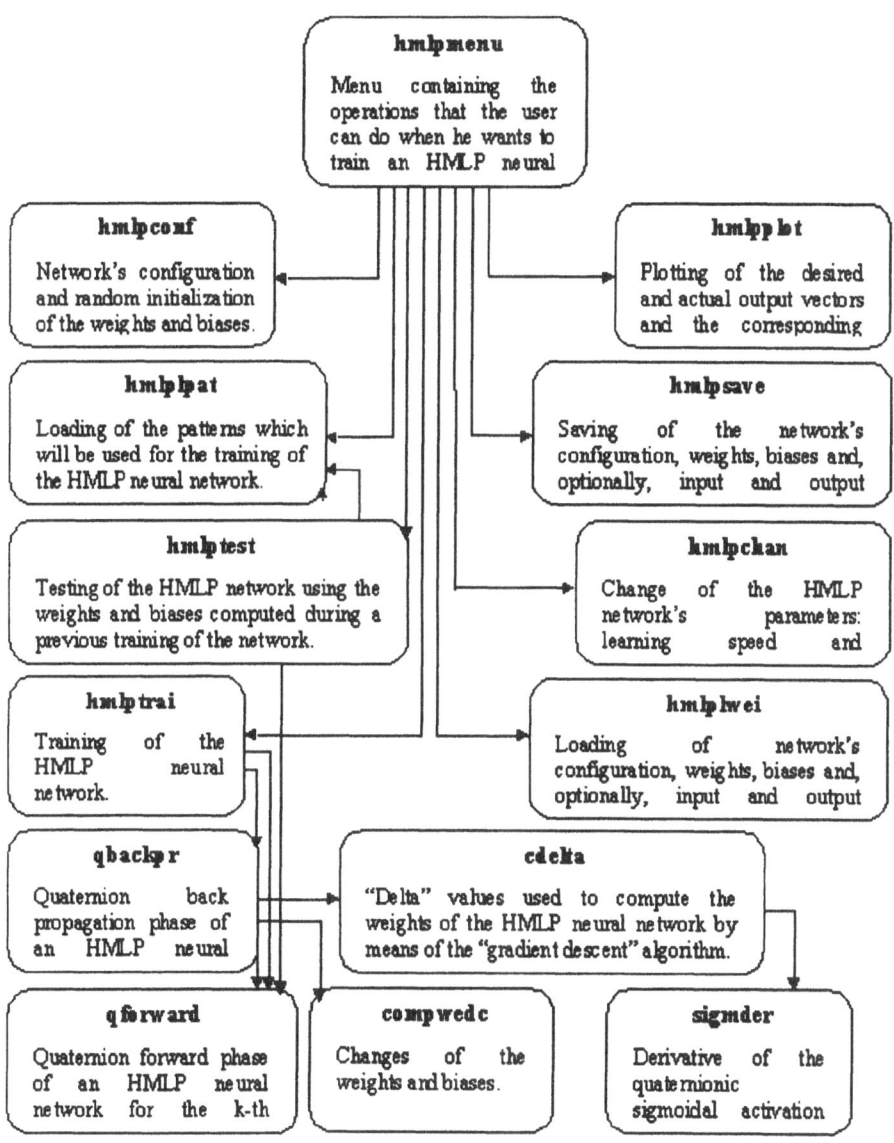

Figure C.1

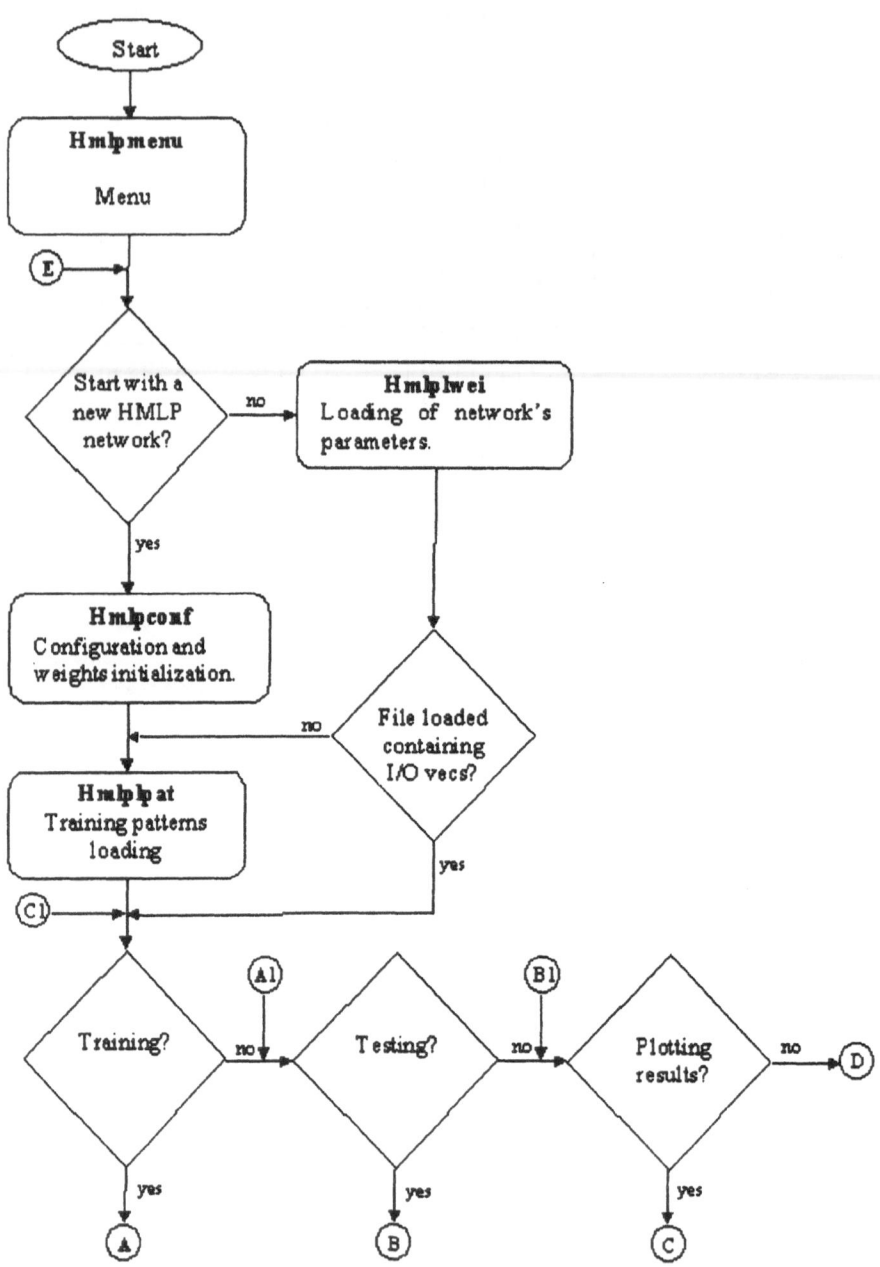

Figure C.2: *Flow Chart*

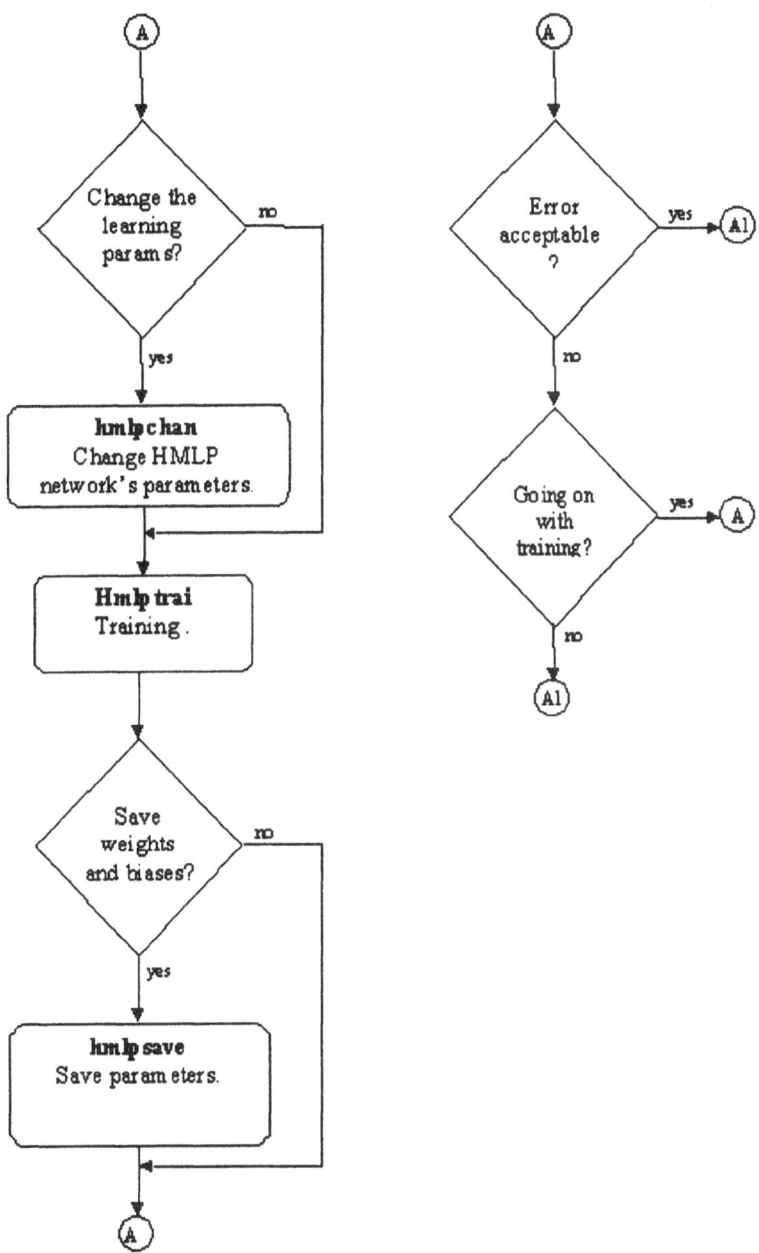

Figure C.3: *Flow Chart*

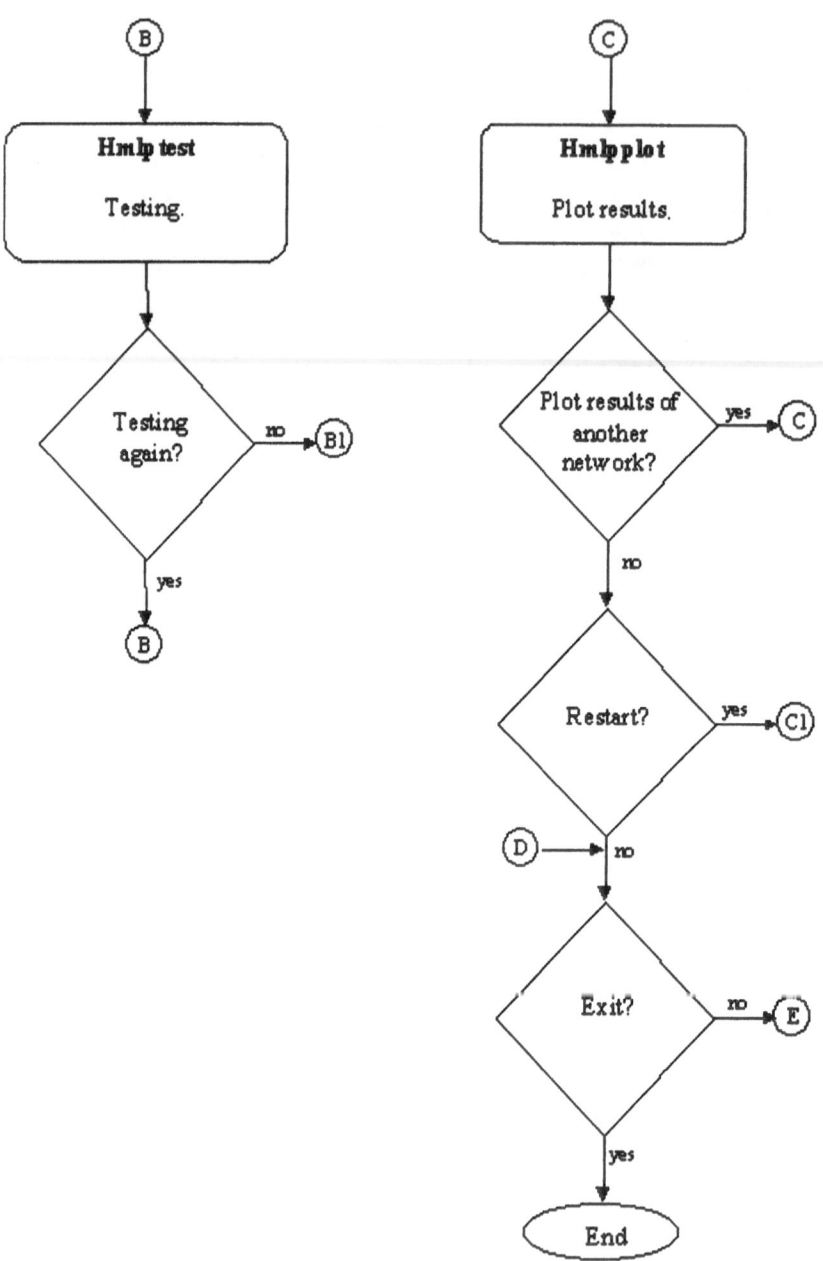

Figure C.4: *Flow Chart*

EXAMPLE 133

C.4 Example

In the following an example of approximation of a generic quaternionic function is reported.

Suppose we want to approximate with an HMLP neural network the following quaternionic function:

$$f(q_1, q_2) = 2(q_1^2 + q2)^2$$

To create the set of patterns, q_1 and q_2 are quaternions initialized with random values belonging to the interval $[0, 1]$ and $f(q_1, q_2)$ is computed. We start with a network with two hidden units. The HMLP network has therefore two input neurons, q_1 and q_2, two hidden neurons and one output neuron, $f(q_1, q_2)$. The network has been trained using 100 input patterns, the first 100 random values of q1 and q2. The first 10 input patterns are the following:

0.2746	0.5250	0.5568	0.4778	0.5192	0.9772	0.7720	0.8791
0.0030	0.4633	0.7390	0.4572	0.7719	0.4445	0.5131	0.3594
0.4143	0.0652	0.3160	0.5200	0.0654	0.0342	0.4867	0.0899
0.0269	0.7134	0.1353	0.4465	0.4428	0.5238	0.6501	0.1610
0.7098	0.4889	0.5285	0.1503	0.9772	0.0316	0.6970	0.4168
0.9379	0.6677	0.3107	0.2859	0.4677	0.7532	0.4075	0.0964
0.2399	0.6820	0.5881	0.8337	0.3291	0.5916	0.4868	0.6747
0.1809	0.1996	0.5181	0.6956	0.4459	0.8229	0.4798	0.5626
0.3175	0.9166	0.4308	0.5778	0.7433	0.3296	0.7105	0.8130
0.8870	0.8659	0.2588	0.6812	0.2053	0.2321	0.5342	0.8782

Each row represents an input pattern while the columns denote the components of the quaternion input neurons, q_1 and q_2. Then the first 4 columns corresponding to the neuron q_1 while the other 4 columns corresponding to the neuron q_2.

The targets corresponding to the 10 input patterns above reported are:

1.1891	2.5057	2.1642	2.2147
1.5439	1.3184	2.1185	1.1368
0.4740	0.0770	1.1731	0.7205
0.8870	2.0655	1.3369	0.7208
2.9620	0.5414	1.9527	0.8788
2.6948	2.3979	1.0082	0.3564
0.7732	2.1136	1.6655	2.7394
0.9573	1.7254	1.4964	2.0930
1.6882	2.3396	1.7922	2.2937
1.9842	1.9637	1.2024	2.6844

The network has been trained with 800 cycles. After the training the HMLP network has been tested, at first using the training patterns and then using different patterns. In both cases the HMLP neural network has well approximated the function. The results have been saved in two files and they are plotted here in Figg. C.5-C.13.

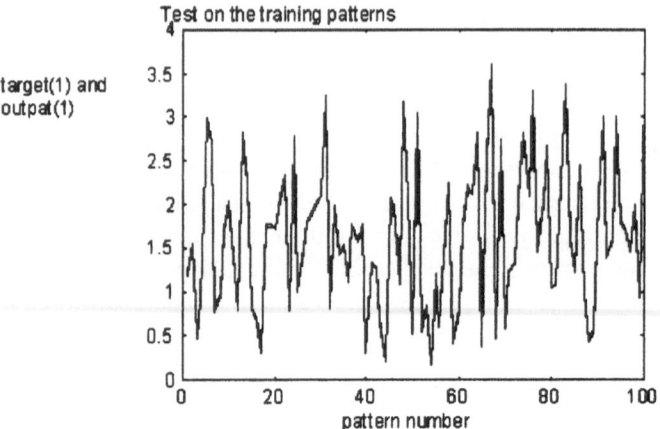

Figure C.5: *Comparison between the first components of the desired and actual outputs computed on the set of training patterns.*

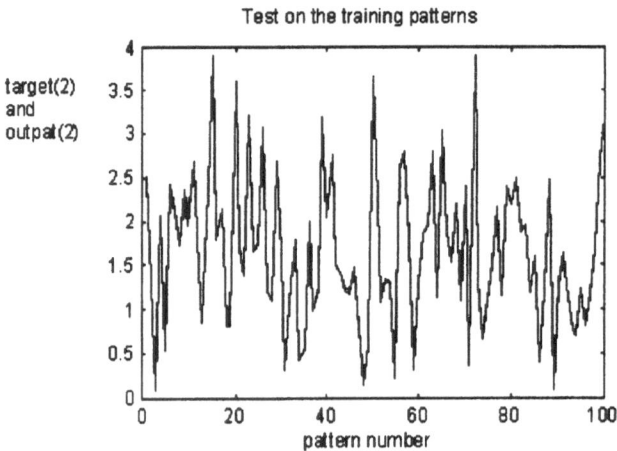

Figure C.6: *Comparison between the second components of the desired and actual outputs computed on the set of training patterns.*

In order to make clearer the difference between the desired and actual outputs, the error relative to the C.5 is reported.
For the training with the HMLP neural networks' program, the following steps have been done:

EXAMPLE 135

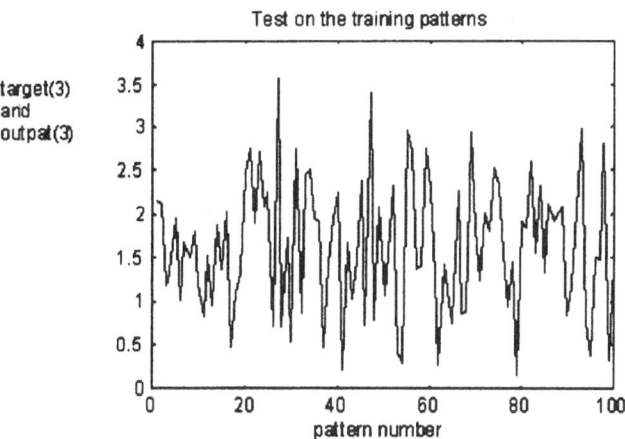

Figure C.7: *Comparison between the third components of the desired and actual outputs computed on the set of training patterns.*

1. configuration of the network:

   - input neurons $= 2$
   - hidden neurons $= 2$
   - output neurons $= 1$;

2. loading of the input patterns and target from a file;

3. choice of the training parameters:

   - learning speed
   - momentum;

4. training of the HMLP neural network with 800 learning cycles;

5. testing of the network;

6. saving of the results in a file;

7. plotting of the results;

8. quit.

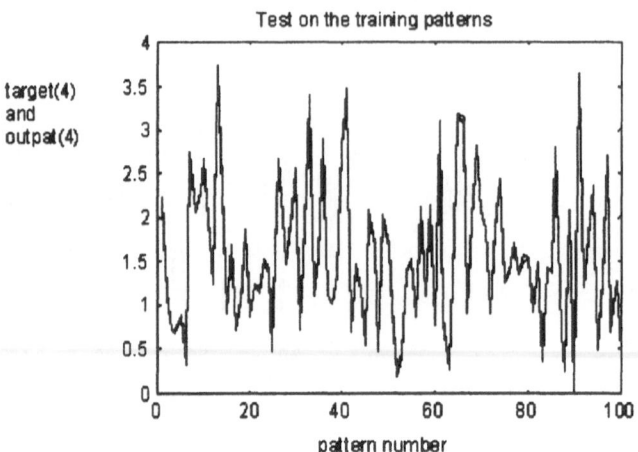

Figure C.8: *Comparison between the fourth components of the desired and actual outputs computed on the set of training patterns.*

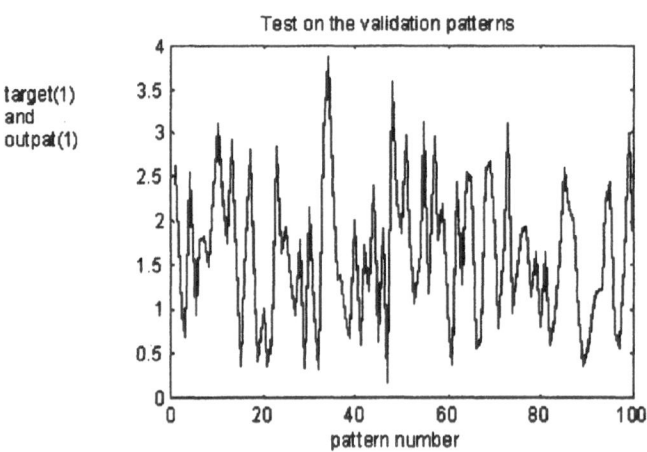

Figure C.9: *Comparison between the first components of the desired and actual outputs computed on a set of testing patterns.*

EXAMPLE 137

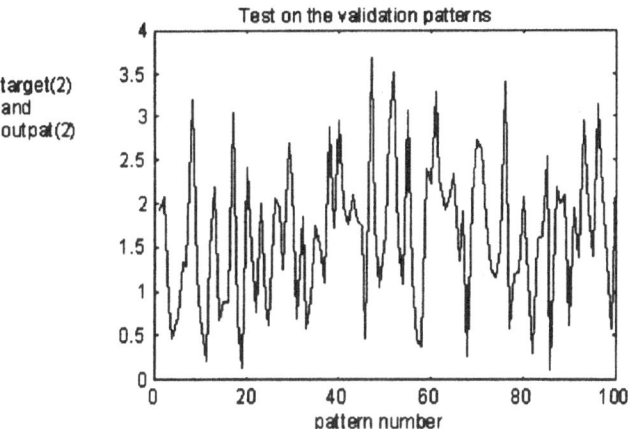

Figure C.10: *Comparison between the second components of the desired and actual outputs computed on a set of testing patterns.*

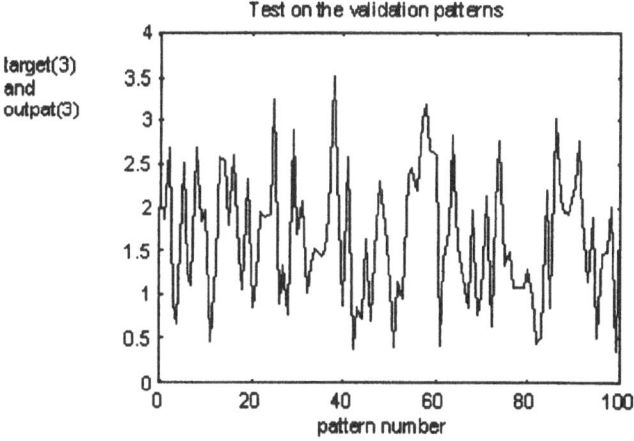

Figure C.11: *Comparison between the third components of the desired and actual outputs computed on a set of testing patterns.*

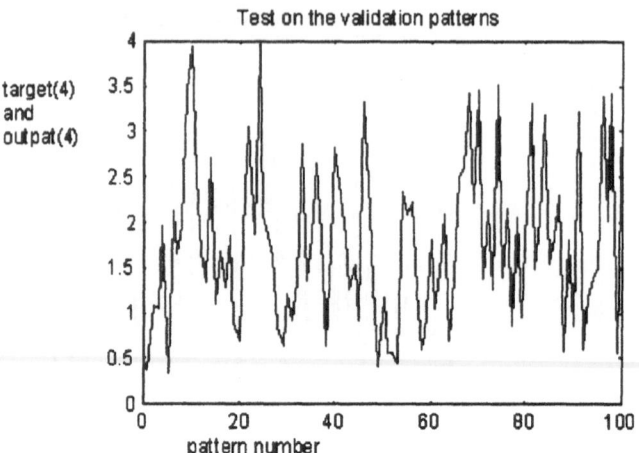

Figure C.12: *Comparison between the fourth components of the desired and actual outputs computed on a set of testing patterns.*

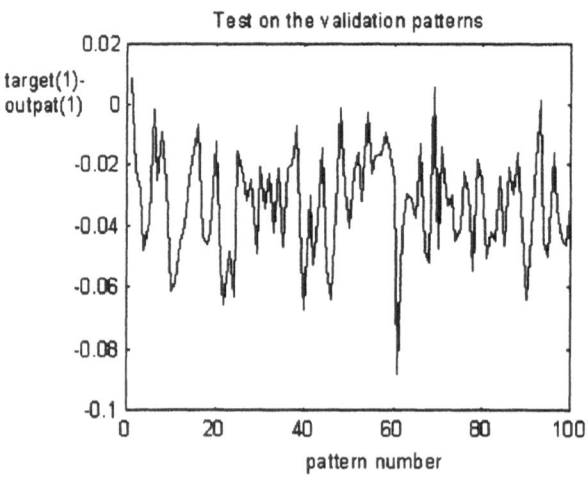

Figure C.13: *Error between the first components of the desired and actual outputs computed for the patterns of fig. C.5*

C.5 HMLP Network

hmlpmenu.m

```
% HMLPMENU
% Menu containing the operations that the user can do when
% he wants to train an HMLP neural network:
%
% 1 - Configuration
% 2 - Loading Patterns
% 3 - Training
% 4 - Saving Weights and Biases
% 5 - Loading Weights and Biases
% 6 - Change Parameters
% 7 - Testing
% 8 - Plotting Results
% 9 - Quit
%
% When the user wants to train an HMLP neural network he
% must first configure the network, then load the patterns
% from a file '.mat' and then train the HMLP network.
% After the configuration of the network it is possible to
% change the training parameters of the network whenever
% the user wants to change  them.
% After the configuration of the HMLP network the user can
% also choose to train the network loading the weights and
% biases from a file rather than initializing them with
% random values.
% It is also possible to load weights, biases, input vector
% and network's configuration from a file without first
% configuring the network.
% During the training of the network it is possible to save
% the weights and biases in a file.
% After the training the user can choose to save only
% the weights and biases. The user must test the HMLP
% network before saving the desired and actual output
% vectors and the input vector, apart from the weights
% and biases.
% Moreover the user can choose to test the network just
% trained using as validation set either that one used for
% training the network or another loading the patterns
% from a different file.
% It is also possible to test a network loading the weights,
```

```
% biases and target from a file.
% The user must decide the number of patterns to use for
% testing the HMLP network.
% He can also choose to plot the results of the HMLP neural
% network's training or testing.
% For each output vector two plots are obtained: one
% compares the desired and actual output vectors and the
% other shows the corresponding error trend.
%
% The input patterns used for training or testing the
% network must form a matrix of column vectors, one for each
% input neuron, called 'inpat'.
% Really each column is a matrix with Np rows (the number of
% patterns) and 4 columns, since the inputs of the HMLP
% neural network are quaternions.
%
%
% Example
%
% Consider an HMLP neural network with 2 input neurons.
% The inputs are the vectors q1 and q2 whose sizes are both
% (Np X 4). The variable of the pattern will be the following:
%
%         inpat = [q1(1) q2(1)]
%
%  Then the size of this variable is (Np X 8).
%
%
% The user must also create the target vector, called 'target',
% that must be a matrix with 4 columns, since the target is a
% quaternion (as the inputs), and (Np*nout) rows, where nout
% is the number of the output neurons.

% Variable used for the 'while' statement.
continue=1;

% Displaying of the menu on the HMLP neural networks.
while(continue==1)

% Variable used for executing a forward phase after the
% network's training.
  train=0;
  clc;
  disp(' ');
```

```
disp(' ');
disp('                    HMLP NEURAL NETWORKS MENU');
disp(' ');
disp(' ');
disp(' ');
disp(' ');
disp(' ');
disp('            1 - Configuration');
disp(' ');
disp('            2 - Loading Patterns');
disp(' ');
disp('            3 - Training');
disp(' ');
disp('            4 - Saving Weights and Biases');
disp(' ');
disp('            5 - Loading Weights and Biases');
disp(' ');
disp('            6 - Change Parameters');
disp(' ');
disp('            7 - Testing');
disp(' ');
disp('            8 - Plotting Results');
disp(' ');
disp('            9 - Quit');
disp(' ');

% Choice of a number from the menu.
choice=input('        Choose a number from the menu: ');

% Execution of the file corresponding to the chosen number.
if choice==1
  eval('hmlpconf');
elseif choice==2
  eval('hmlplpat');
elseif choice==3
  eval('hmlptrai');
elseif choice==4
  eval('hmlpsave');
elseif choice==5
  eval('hmlplwei');
elseif choice==6
  eval('hmlpchan');
elseif choice==7
  eval('hmlptest');
```

```
  elseif choice==8
    eval('hmlpplot');
  elseif choice==9
    break;
  end

end

disp(' ');
disp(' ');

hmlpconf.m
```

```
% HMLPCONF
% Introduction of the HMLP neural network's configuration and
% random initialization of the weights and biases.



% Clear all variables.
clear all;

% Setting of the variable used for the 'while' statement
% contained in the file
% 'hmlpmenu.m'.
continue=1;

% Clear command window.
clc;

disp(' ');
disp(' ');
disp('             CONFIGURATION OF THE HMLP NETWORK');
disp(' ');
disp(' ');

% Introduction of the number of the input, hidden and output
% neurons.
ninp=input('  Insert number of input neurons:  ');
nhid=input('  Insert number of hidden neurons: ');
nout=input('  Insert number of output neurons: ');

% Computation of the number of weights of the network.
nwei=ninp*nhid+nout*nhid;
```

```
% Random initialization of the weights and the biases of the
% HMLP neural network.
weight=rand(4,nwei);
bias=rand(4,nhid+nout);

% Setting of the HMLP network's parameters with default
% values.
eps=0.2;
momentum=0.01;

% Initialization of the effected total cycle number.
t_cy_num=0;

hmlplpat.m


% HMLPLPAT
% Loading of the patterns which will be used for the training
% of the HMLP neural network.

% Clear command window.
clc

disp(' ');
disp(' ');
disp('           LOADING OF THE PATTERNS FROM A FILE');
disp(' ');
disp(' ');

% Loading of the file which contains the pattern vector and
% the target vector.
infile=input('  Insert the name of the pattern file: ','s');
eval(['load ',infile,';']);

hmlptrai.m


% HMLPTRAI
% Training of the HMLP neural network.

% Declaration as global of the variables used in 'qbackpr1'
% but setted here.
 global momentum dweight dbias

% Initialization of the variables which will contain the
```

```
% previous changes of the weights and biases.
dweight=0;
dbias=0;

% Clear command window.
clc

disp(' ');
disp(' ');
disp('              TRAINING OF THE HMLP NEURAL NETWORK');
disp(' ');
disp(' ');

% Introduction of the number of the learning cycles used
% during the learning
% phase of the neural network.
disp(' ');
ci=input('   Insert number of learning cycles: ');

disp(' ');
ncsave=input('   How often cycles do you want to save
the weights and biases
                         (0 = no saving)? ');
if ncsave>0
  savfile=input('   Insert the name of the file in which
 saving weights and biases:',

end


% Introduction of the number of patterns that the user
% wants to use for training the HMLP neural network.
Np=input('   Insert the number of patterns you want to use
 to train the HMLP network: ');

% Testing if the variable 'inpat' exists.
if exist('inpat')==1

  % Clear command window.
  clc

% Counter of the number of the learning cycles before
% saving the HMLP network's weights and biases.
  contsave=0;
```

```
% Construction of the target vector containing Np elements
% for each output neuron.
targetd=[];
Ninpat=size(inpat,1);
for xx=1:nout
targetd=[targetd;target(((xx-1)*Ninpat+1):((xx-1)*Ninpat+Np),:)];
end

disp(' ');

for hh=1:ci        % learning cycles

  % Compute the total number of learning cycles.
  t_cy_num=t_cy_num+1;

  contsave=contsave+1;

  % Display the number of the learning cycles just done.
  visual=['cycle_number = ' int2str(hh) '
     (total_cycle_number = '
               int2str(t_cy_num) ')'];
  disp(visual);


  % Back propagation phase of the network for Np patterns.
[weight,bias]=qbackpr(inpat(1:Np,:),targetd,weight,bias,nhid,
ninp,nout);

% Saving of the HMLP network's weights and biases every
% 'ncsave' learning cycles.
    if contsave==ncsave
eval(['save ',savfile,'.mat t_cy_num weight bias ninp
nhid nout eps momentum;' ]);
      contsave=0;
    end

  end

  % Removal of the variable 'outpat' from the workspace
 because its value is
  % old.
  if exist('outpat')==1
    clear outpat;
```

```
   end

   % Forward phase on all the patterns of the HMLP neural
   % network.
   train=1;
   eval('hmlptest');

else

   disp(' ');
   disp(' ');
   disp('   The variable <<inpat>> does not exist.');
   disp('   You must first load the patterns');

   % Pause of 2 seconds.
   pause(2);

end

% Pause of 2 seconds.
pause(2);

hmlpsave.m


% HMLPSAVE
% The HMLP network's configuration and the weights and biases
% obtained from the training of the network are saved in a
% file (in the current directory).

% Clear command window.
clc

disp(' ');
disp(' ');
disp('          SAVING OF THE WEIGHTS, BIASES, INPUT AND
               OUTPUT VECTORS IN A FILE');
disp(' ');
disp(' ');

% Testing if the variable 'outpat' exists.
if exist('outpat')==0

savfile=input('   Insert the name of the file in which
 saving weights and biases: '
```

```
eval(['save ',savfile,'.mat t_cy_num weight bias
ninp nhid nout eps momentum;'

else
  risp=input('  Do you want to save only <w>eights and
 biases or also input and output <v>ectors? ','s');
  savfile=input('  Insert the name of the file in which
 saving data: ','s');
  if risp=='w'
eval(['save ',savfile,'.mat t_cy_num weight bias ninp nhid
nout eps momentum;' ]);
  elseif risp=='v'
    eval(['save ',savfile,'.mat inpat outpat target t_cy_num
weight bias ninp nhid nout eps momentum;' ]);
  end
end
```

hmlplwei.m


```
% HMLPLWEI
% Loading of the weights and biases which will be used for
% the training of the HMLP neural network.

% Clear command window.
clc

disp(' ');
disp(' ');
disp('        LOADING OF THE WEIGHTS AND BIASES FROM A
             FILE');
disp(' ');
disp(' ');

% Loading of the file which contains the weights and biases
% of the previously
% trained HMLP network.
wbfile=input('  Insert the name of the file containing
weights and biases: ','s');
eval(['load ',wbfile,';']);
```

hmlpchan.m

```
% HMLPCHAN
% Change of the HMLP network's parameters:
% - learning speed
% - momemtum.


% Clear command window.
clc

disp(' ');
disp(' ');
disp('            CHANGE OF THE VALUES OF THE LEARNING
                      SPEED AND THE MOMENTUM');
disp(' ');
disp(' ');

% Displaying of the previous values of the parameters.
disp('   The previous values of the parameters are:');
disp(' ');

% Displaying of the 'eps' value.
if eps==0 disp(['        eps      =  ' int2str(eps) '.0']);
else
  strdisp=['        eps      =  ' int2str(floor(eps)) '.'];
  epsdisp=(eps-floor(eps))*10;
  while(epsdisp>0)
    strdisp=[strdisp int2str(floor(epsdisp))];
    epsdisp=(epsdisp-floor(epsdisp))*10;
  end
  disp(strdisp);
end

% Displaying of the 'momentum' value.
if momentum==0 disp(['momentum  =  ' int2str(momentum) '.0']);
else
strdisp1=['   momentum  =  ' int2str(floor(momentum)) '.'];
  momdisp=(momentum-floor(momentum))*10;
  while(momdisp>0)
    strdisp1=[strdisp1 int2str(floor(momdisp))];
    momdisp=(momdisp-floor(momdisp))*10;
  end
  disp(strdisp1);
end
```

```
disp(' ');

% Testing if the user wants to change the training parameters.
risp=input('  Do you want to change the training parameters?
(y/n) ','s'); if risp=='y'

  % Introduction of the learning speed used to update weights
% and biases.
  eps=input('  Insert learning speed: ');
  momentum=input('  Insert momentum:      ');

end

hmlptest.m


% HMLPTEST
% Testing of the HMLP network using the weights and biases
% just computed for the patterns of the training set.


% Control whether the testing is required by the user or is
% only the forward phase on all the patterns after the
% training of the HMLP network.
if train==0

  % Clear command window.
  clc

  disp(' ');
  disp(' ');
  disp('          TESTING OF THE HMLP NEURAL NETWORK');
  disp(' ');
  disp(' ');

  % Loading of the testing patterns if they differ from the
  % training ones.
  risp=input('  Do you want to test the HMLP network on
  the training patterns?
                    (y/n) ','s');
  if risp=='n'
    eval('hmlplpat');
    disp(' ');
```

```
    end

    % Introduction of the number of patterns that the user
    % wants to use for
    % testing the HMLP neural network.
    numpat=input('   Insert the number of patterns you want
  to use for testing the HMLP network: ');




  else

    % Number of patterns that the user wants to use for testing
    % the HMLP neural network.
    numpat=Np;

  end

  % Construction of the input vector used in the function
  % 'qforward'.
  inpp=[];
  for kk=1:numpat
    for nn=1:ninp
      inpp=[inpp inpat(kk,((nn-1)*4+1):(nn*4))'];
    end
  end

  % Forward phase of the network; 'outpat' is the output vector
  % of the HMLP network.
  tt=0;
  outpat=zeros(4,(numpat*nout));
  for k=1:ninp:(ninp*numpat)
    [actv,netv,outv]=qforward(ninp,nhid,nout,weight,bias,
     inpp(:,k:(k+ninp-1)));
    tt=tt+1;
    for i=1:nout
      outpat(:,(tt+numpat*(i-1)))=outv(:,i);
    end
  end
  outpat=outpat';

  % Compute the square error only if this procedure is
  % required by the user.
```

```
if train==0

  % Length of the desired output vector.
  x=size(target,1)/nout;

  % Square error on all patterns.
  err=[];
  for s=1:nout
    errq=target(((s-1)*x+1):((s-1)*x+numpat),
    :)-outpat(((s-1)*numpat+1):(s*numpat),:);
    err=[err;errq];
  end

  disp(' ');
  quadr_err=sum(normvett(err))

  % Pause of 2 seconds.
  pause(2);

end

hmlpplot.m
% HMLPPLOT
% The output vector of the HMLP network is plotted and it
% is compared with
% the desired output vector. Moreover the error between
% the desired and actual output vectors is plotted.

% Clear command window.
clc

disp(' ');
disp(' ');
disp('    PLOTTING OF THE OUTPUT VECTORS AND THE
          CORRESPONDING ERROR VECTOR');
disp(' ');
disp(' ');

% Choice of the output vectors that the user wants to plot.
risp=input(' Do you want to plot the <c>urrent output
vectors or <o>thers (loading them from a file)? ','s');

if risp=='o'
```

```
% Loading of the file which contains the desired and
% actual output vectors.
  outfile=input('   Insert the name of the file containing
the output vectors you want to plot: ','s');
  eval(['load ',outfile,';']);


end

% Length of the desired output vector.
x=size(target,1)/nout;

% Length of the actual output vector.
x1=size(outpat,1)/nout;

% Vector representing the 'x' axis used for plotting
% the desired and actual
% output vectors and the error between them.
npat=1:x1;

for s=1:nout

  for i=1:4

% Plot of the i-th components of the s-th desired and actual output
% vectors.
  plot(npat,target(((s-1)*x+1):((s-1)*x+x1),i),npat,
  outpat(((s-1)*x1+1):(s*x1),i));
  title(['target' int2str(s) '(' int2str(i) ')   and
 outpat' int2str(s) '(' int2str(i) ')']);

    % Pause of 2 seconds.
    pause(2);
    % Command for creating and opening a new figure.
    figure;

    % Plot of the i-th component of the s-th error vector
    % between the desired and actual output vectors.
    plot(npat,target(((s-1)*x+1):
    ((s-1)*x+x1),i)-outpat(((s-1)*x1+1):(s*x1),i));
    title(['error' int2str(s) '(' int2str(i) ')']);

    if i<4
      % Pause of 2 seconds.
      pause(2);
```

```
      % Command for creating and opening a new figure.
      figure;
    end

  end

  if s<nout
    % Pause of 2 seconds.
    pause(2);
    % Command for creating and opening a new figure.
    figure;
  end
end
```

qforward.m

```
function[act,net,outv]=qforward(ninp,nhid,nout,weight,bias,inp)
%QFORWARD Quaternion forward phase of a HMLP network for the
%k-th pattern.
%
% The syntax of the function QFORWARD is:
% [ACT,NET,OUTV]=QFORWARD(NINP,NHID,NOUT,
%    WEIGHT,BIAS,INP)
%
% NINP is the input neuron number. NHID is the hidden
%  neuron number. NOUT is the output neuron number.
% WEIGHT is the weight matrix. BIAS is the bias matrix.
% INP is the input vector.
%
% ACT is the vector of the neurons' activation functions.
% NET is the activation value. OUTV is the k-th element of
% the HMLP output vector used for the system control.


% Initialization of the output vector 'outv' returned
% by this function.
outv=[];

% Initialize the matrix of the activation functions.
act=zeros(4,nhid+ninp+nout);

% Compute the activation functions of all input neurons.
act(:,1:ninp)=inp;
```

```matlab
t=1;

% Initialize the matrix of the activation values.
net=zeros(4,nhid+nout);

for i=1:nhid


  % Compute the activation value of the i-th hidden neuron.
  net(:,i)=bias(:,i);
  for j=1:ninp
    net(:,i)=net(:,i)+prodquat(weight(:,t),act(:,j));
    t=t+1;
  end

  % Compute the activation function of the i-th hidden neuron.
  for s=1:4
    act(s,i+ninp)=1/(1+exp(-net(s,i)));
  end

end

for i=1:nout

  % Compute the activation value of the i-th output neuron.
  net(:,i+nhid)=bias(:,i+nhid);
  for j=1:nhid
  net(:,i+nhid)=net(:,i+nhid)+prodquat(weight(:,t),
  act(:,j+ninp));
    t=t+1;
  end

% Compute the activation function of the i-th output neuron.
  act(:,i+ninp+nhid)=net(:,i+nhid);

  % Compute the output vector of the HMLP neural network.
  outv=[outv act(:,i+ninp+nhid)];

end

qbackpr.m


function [weight,bias]=qbackpr(inpat,target,weight,bias,nhid,
```

```
                 ninp,nout)
%QBACKPR Quaternion back propagation phase of a HMLP network.
%
% The syntax of the function QBACKPR is:
% [WEIGHT,BIAS]=QBACKPR(INPAT,TARGET,WEIGHT,
%      BIAS,NHID,NINP,NOUT)
%
% INPAT is the input vector.TARGET is the vector of the
% desired output of the system. WEIGHT is the weights
% matrix. BIAS is the bias matrix. NHID is the hidden
% neurons number. NINP is the input neurons number.
% NOUT is the output neurons number.
%
% At the output WEIGHT is the modified weight matrix
% and BIAS is the modified bias matrix.


% Declaration as global of the variables defined in 'hmlptrai.m' and
% 'hmlpconf.m'.
global momentum dweight dbias

% Size of the input vector (number of patterns).
[x,y]=size(inpat);

%**************** Forward phase **********************

tse=0;

for kkk=1:x

  % Extraction of the kkk-th pattern from the input vector.
  inp=[];
  for nn=1:ninp
    inp=[inp inpat(kkk,((nn-1)*4+1):(nn*4))'];
  end

  wedt=0;
  bedt=0;

  % Forward phase for the kkk-th input pattern.
  [act,net,out]=qforward(ninp,nhid,nout,weight,bias,inp);

  % Computation of the kkk-th value of the desired output.
  target1=[];
```

```
  for kkkout=1:nout
    target1=[target1;target(((kkkout-1)*x+kkk),:)];
  end

  % Error between the desired and actual outputs of the
  % HMLP neural network.
  err=target1'-out;

  % Square error on the kkk-th pattern.
  pse=0;
  for s=1:nout
    pse=pse+normq(err(:,s));
  end

  % Computation of the square error on all the patterns.
  tse=tse+pse;

% *********** Back-propagation phase ******************

  % Computation of the delta values used to update weights
  % and biases.
  [delta]=cdelta(net,weight,err,ninp,nhid,nout);

  % Computation of the weights and biases' changes for the
  % kkk-th pattern.
  [wed,bed]=compwedc(act,delta,ninp,nhid,nout);

  % Computation of the weights and biases' changes for
  % all the patterns.
  wedt=wedt+wed;
  bedt=bedt+bed;

  % On-line updating of weights and biases.
  dweight=eps*wedt+momentum*dweight;
  weight=weight+dweight;
  dbias=eps*bedt+momentum*dbias;
  bias=bias+dbias;

end

% Displaying of the scalar component of the square error.
square_error=tse(1)


cdelta.m
```

```
function delta=cdelta(net,weight,err,ninp,nhid,nout);
%CDELTA "Delta" values used to compute the weights of the
% neural network by means of the "gradient descent" algorithm.
%
%CDELTA(NET,WEIGHT,ERR,NINP NHID,NOUT) returns the
%"delta" values.
%
%NET is the activation value. WEIGHT is the weight matrix.
%ERR is the difference between the desired output and the
%actual output of a neuron. NINP is the input neuron number.
%NHID is the hidden neuron number. NOUT is the output
%neuron number.


% Initialization of the matrixes which contain, respectively,
% the "delta" values and the errors of the neural network's
% neurons.
delta=zeros(4,nout+nhid);
errv=zeros(4,nout+nhid);

for i=1:nout

  % Error signal of each output neuron.
  errv(:,i+nhid)=err(:,i);

  % "Delta" value of each output neuron.
  delta(:,i+nhid)=err(:,i);

end

t=ninp*nhid;

for i=1:nout

  % Computation of the error signal of each hidden neuron.
  for j=1:nhid
    t=t+1;
    a=prodquat(quacon(weight(:,t)),delta(:,i+nhid));
    errv(:,j)=errv(:,j)+a;
  end

end
```

```
for i=1:nhid

  % Computation of the "delta" value of each hidden neuron.
  delta(:,i)=quaprodc(errv(:,i),sigmder(net(:,i)));

end
```

compwedc.m

```
function [wedc,bedc]=compwedc(actc,deltac,ninpc,nhidc,noutc);
%COMPWEDC  Changes of the weights and the biases.
%
% The syntax of the function COMPWEDC is:
% [WEDC,BEDC]=COMPWEDC(ACTC,DELTAC,NINPC,
%           NHIDC,NOUTC)
%
%ACTC is the vector of the neurons' activation functions.
%DELTAC is the vector of the "delta" values used to compute
%the weights and the biases of the neural network. NINPC
%is the input neuron number. NHIDC is the hidden neuron
%number. NOUTC is the output neuron number.
%
%WEDC and BEDC are the vectors used to compute the
%changes of weights and biases.
%WEDC and BEDC must be multiplied by the learning speed
%to obtain the changes of weights and biases.


t=1;

for i=1:nhidc

  %Change of the weights between input and hidden neurons.
  for j=1:ninpc
    wedc(:,t)=prodquat(deltac(:,i),quacon(actc(:,j)));
    t=t+1;
  end

  %Change of the biases of the hidden neurons.
  bedc(:,i)=deltac(:,i);

end
```

```
for i=1:noutc

  %Change of the weights between hidden and output neurons.
  for j=1:nhidc
wedc(:,t)=prodquat(deltac(:,i+nhidc),quacon(actc(:,j+ninpc)));
    t=t+1;
  end

  %Change of the biases of the output neurons.
  bedc(:,i+nhidc)=deltac(:,i+nhidc);
end
```

sigmder.m

```
function [q]=sigmder(netu)
%SIGMDER  Derivative of the quaternionic activation function.
%
%SIGMDER(NETU) returns the derivative of the quaternionic
%activation function whose expression is:
%QF = 1 / (1 + exp(-NETU))
%NETU is the activation value.


for i=1:4

 %Activation function
 sq(i,1)=1/(1+exp(-netu(i)));

 %Derivative of the activation function
 q(i,1)=sq(i,1)*(1-sq(i,1));

end
```

References

[1] *Signal Processing for Control*, Lecture Notes in Control and Information Sciences, N 79, Edited by K. Godfrey, P. Jones.

[2] T. Nitta, *A Back-Propagation Algorithm for Complex Numbered Neural Networks*, Proc. of 1993 Int. Joint Conference on Neural Networks, 1993.

[3] G. Georgiou, C. Koutsougeras, *Complex Domain Back-Propagation* , IEEE Trans. on Circuits and Systems-II: Analog and Digital Signal Processing, vol. 39, No. 5, pp. 330-334, May 1992.

[4] N. Benvenuto, M Marchesi, F. Piazza, A. Uncini , *A Comparison Between Real and Complex valued Neural Networks in communication areas*, ICANN-91, Espoo, Finland, pp. 1177-1180, June 1991.

[5] N. Benvenuto, M Marchesi, F. Piazza, A. Uncini , *Nonlinear satellite Radio Links Equalized Using Blind Neural Networks*, Proc. IEEE Int. Conf. on Acoustic, Speech and Signal Processing, ICASSP '91, Toronto, Canada, 1991.

[6] P. Arena, L. Fortuna, M. G. Xibilia, *Predicting complex chaotic time series via complex valued MLPs*, Proc. of IEEE Int. Conference on Circuits and Systems, pp. 29-32, London, 1994.

[7] R. M. Range *Holomorfic functions and integral representations in several complex variables*, Graduate text in Mathematics, Springer-Verlag, 1986.

[8] W. Rudin *Real and complex analysis*, McGraw-Hill, New York, 1966.

[9] E. Giusti, *Analisi matematica 2*, Boringhieri 1983.

[10] M.S. Kim, C.C. Guest, *Modification of back-propagation for complex-valued signal processing in frequency domain*, Proc. of Int. Joint Conf. on Neural Networks, pp. 27-31, San Diego, U.S.A, June 1990.

[11] *H. Leung and S. Haykin, The complex Backpropagation algorithm, IEEE Trans. On Signal Processing, Vol.39, No.9, pp. 2101-2104, Sept. 1991.*

[12] N. Benvenuto, F. Piazza, On the complex Backpropagation algorithm, IEEE Trans. On Signal Processing, Vol.40, No.4, April 1992.

[13] D. E. Rumelhart, McClelland, Parallel Distributed Processing: Exploration in the microstructure of cognition, pp. 318-362, MIT Press, 1986.

[14] B.Widrow, J. Mc Cool, M. Ball, The complex LMS algorithm Proc. of the IEEE, pp. 719-720, April 1975.

[15] G. Cybenko, Approximation by superposition of a sigmoidal function, Mathematics of control signals and systems, Vol.2, pp. 303-314, 1989.

[16] A. N.Kolmogorov, On the representation of continuous functions of many variables by superposition of continuous functions of one variable and addition, Doklady Akademii Nauk SSSR, 144, 679-681; American Mathematical Society Translation, 28, pp. 55-59, 1963.

[17] D. A. Sprecher, On the structure of continuous functions of several variables, Trans. of the American Mathematical Society, 115, pp.340-355, 1965.

[18] R. Heicht-Nielsen, Neurocomputing, Addison-Wesley Publishing Company, 1991.

[19] K.I. Funahashi, On the approximate realization of continuous mappings by Neural Networks, Neural Networks, Vol.2, pp.183-192, 1989.

[20] K. Hornik, M. Stinchcombe and H. White, Mulilayer Feedforward Networks are universal approximators, Neural Networks, Vol.2, pp.359-366, 1989.

[21] W. Rudin, Functional Analysis, MCGraw-Hill series in higher Mathematics, 1973.

[22] P. Arena, L. Fortuna, R. Re, M. G. Xibilia, On the capability of neural networks with complex neurons in complex valued function approximation, Proc. IEEE Int. Conf. on Circuits and Systems, pp. 2168-2171, Chicago, 1993.

[23] P. Arena, Reti neurali per modellistica e predizione in circuiti e sistemi, PhD Dissertation, University of Catania, 1994.

[24] K. Ikeda, H. Daido, O. Akimoto, Optical Turbolence: Chaotic behavior of transmitted light from a Ring Cavity, Physical Review Letters, Vol.45, No.9, pp.709-712, 1980.

[25] X. Cui, K. Shin, Intelligent Coordination of Multiple Systems with Neural Networks, IEEE Trans. on System, Man and Cybernetic, Vol. 21, No. 6 , pp. 1488-1496, 1991.

[26] T. Nitta, *A Back-Propagation Algorithm for Neural Networks Based on 3D Vector Product*, Proc. of 1993 Int. Joint Conference on Neural Networks, pp.589-592, Portland, 1993.

[27] W. R. Hamilton, *Lectures on Quaternions*, Hodges and Smith, Dublin, 1853.

[28] W. R. Hamilton, *Elements of Quaternions*, 3rd edition, Chelsea, New York, 1969.

[29] Ebbinghaus et al., *Numbers*, G.T.M., Springer Verlag, 1990.

[30] K. Guerlebeck, Wolfgang Sproessig, *Quaternionic Analysis and Elliptic Boundary Value Problems*, Int. Series of Numerical Mathematics, vol. 89, Birkhauser.

[31] C. A. Deavours, *The quaternionic calculus*, Am. Math. Monthly 80, pag. 995-1008,1973.

[32] W. Sproessig, K. Guerlebeck, *An Hypercomplex Method of Calculating Stresses in 3-dimensional Bodies*, Rend, Circ. Mat. Palermo, (2), 6, pag.271-284, 1984.

[33] J. C. K. Chou *Quaternion Kinematic and Dynamic Differential Equation*, IEEE Trans. on Robotics and Automation, Vol. 8, N. 1, pp. 53-64, Feb. 1992.

[34] R. H. Taylor *Planning and Execution of Straight Line Manipulator Trajectories*, IBM J. Research Development, Vol. 23, N. 4, pp. 424-436, July 1979.

[35] K. W. Spring, *Euler parameters and the use of quaternion algebra in the manipulation of finite rotations: a review*, Mechanism Machine Theory, vol. 21, n.5, pp.365-373, 1986.

[36] J. Chou, M. Kamel, *Finding the position and orientation of a sensor on a robot manipulator using quaternions*, Int. J. Robotics Res., vol 10, n. 3, pp.240-254, June 1991.

[37] K. Ueda, S. Takahashi, *Digital filters with hypercomplex coefficients*, Proc. of IEEE Int. Symp. on Circuit and Systems, pp. 479-482, 1993.

[38] J. Kelley, *General Topology*, G.I.M. Springer Verlag, 1975.

[39] P. Arena, L. Fortuna, L. Occhipinti, M.G. Xibilia, *Neural Networks for quaternion valued function approximation*, Proc. of 1994 IEEE Int. Symp. on Circuit and Systems, vol. 6, pp.307-310, London, 1994.

[40] L. Ljung *System identification: theory for the user*, Prentice Hall Information and System Science Series, 1987.

[41] K. S. Narendra, K. Parthasarathy, *Identification and control of dynamical systems using neural networks, IEEE Trans. On Neural Networks, Vol.1 No.1, pp. 4-27, 1990.*

[42] J.Sjoberg, H. Hjalmarsson, L. Ljung, *Neural Networks in System Identification, Proc. of 10th IFAC Symp. on System Identification, Copenaghen, 1994.*

[43] T. Parker and L. O. Chua, *Chaos: A tutorial for Engineers, Proc. of The IEEE, Vol.75, No.8, August 1987.*

[44] A. Lapedes and R. Farber, *Nonlinear signal processing using neural networks: prediction and system modelling, LA-UR-87-2662, 1987.*

[45] A. Lapedes and R. Farber, *How neural nets work, IEEE Denver Conf. on Neural Nets, 1987, preprints.*

[46] H. English, Y. Hiemstra, *The Correlation as Cost Function in Neural Networks, Proc. of IEEE WCCI, pag. 1370-1372, Orlando, 1994.*

[47] E.N.Lorentz, *Deterministic Nonperiodic Flow, Journal Atmospheric Science, 20, 130, 1963.*

[48] L.O.Chua, *Chua's Circuit: Ten Years Later, 1993 International Symposium on Nonlinear Theory and its Applications, Hawaii, Dec 1993.*

[49] K. Mitsubori, T. Saito, *Torus Doubling and Hyperchaos in a Five Dimensional Hysteresis Circuit, Proc. of 1994 IEEE Int. Symp. on Circuit and Systems, vol. 6, pp. 113-116, London.*

[50] S. Lang, *Linear Algebra Addison Wesley Publ. Comp., 1966.*

[51] S.M. Joshi, A.G. Kelkar, and J.T. Wen, *Robust Attitude Stabilization of Spacecraft Using Nonlinear Quaternion Feedback IEEE Trans. on Automatic Control, Vol. 40, N. 10, pp. 1800-1803, Oct. 1995.*

[52] W. T. Miller, R. S. Sutton, P.J. Werbos (edited by) *Neural Networks for Control, A Bradford Book, The MIT Press, Cambridge, 1991.*

[53] J. Wen, K. Delgado, *The attitude control problem, IEEE Trans. on Robotics and Automation, vol. 8, n. 1, pp. 1148-1162, 1992.*

[54] R. Paul, *Robot manipulators:Mathematics, Programming and Control MIT Press, 1983.*

[55] T. Dwyer, *Exact nonlinear control of large angle rotational maneuvers, IEEE Trans. on automatic control, vol. AC 29, n. 9, pp. 769-774, Sept. 1984.*

[56] D. Nguyen, B. Widrow, Neural Networks for Self-Learning Control Systems, IEEE Control System Magazine, pp. 308-313, April 1990.

[57] L. Fortuna, G. Muscato, M.G. Xibilia, An hypercomplex neural platform for neural positioning, IEEE Int. Symposium on Circuit and System, pp.609-612, Atlanta, U.S.A., May 1996.

[58] W. Rudin, Principles of Mathematical analysis, McGraw-Hill, New York, 1964.

[59] S. Baglio, Comportamenti non lineari e fenomeni caotici nei circuiti e nei sistemi dinamici, PhD Dissertation, University of Catania, 1994.

[60] M.F.Barnsley, Fractals Everywhere, Academic Press Professional, 1993.

[61] P. Arena, L. Fortuna, M.G. Xibilia, Neural networks in multidimensional domains, Journal of System Eng., vol.6, p.1-11, 1996.

[62] P. Arena, L. Fortuna, S. Graziani, G. Nunnari A Monitoring approach for the design of Multi layer Neural Networks, Proc. COMADEM 91, Southampton, U.K., 1991

[63] R. Katayama, Y. Kajtani, K. Kuvuta, Y. Nishda, Self generating radial basis function as neuro-fuzzy model and its application to non linear prediction of chaotic time series, Second IEEE Int. Conf. on Fuzzy Systems, San Francisco,USA, 1993.

Lecture Notes in Control and Information Sciences

Edited by M. Thoma

1993–1998 Published Titles:

Vol. 186: Sreenath, N.
Systems Representation of Global Climate
Change Models. Foundation for a Systems
Science Approach.
288 pp. 1993 [3-540-19824-5]

Vol. 187: Morecki, A.; Bianchi, G.;
Jaworeck, K. (Eds)
RoManSy 9: Proceedings of the Ninth
CISM-IFToMM Symposium on Theory and
Practice of Robots and Manipulators.
476 pp. 1993 [3-540-19834-2]

Vol. 188: Naidu, D. Subbaram
Aeroassisted Orbital Transfer: Guidance
and Control Strategies
192 pp. 1993 [3-540-19819-9]

Vol. 189: Ilchmann, A.
Non-Identifier-Based High-Gain Adaptive
Control
220 pp. 1993 [3-540-19845-8]

Vol. 190: Chatila, R.; Hirzinger, G. (Eds)
Experimental Robotics II: The 2nd
International Symposium, Toulouse,
France, June 25-27 1991
580 pp. 1993 [3-540-19851-2]

Vol. 191: Blondel, V.
Simultaneous Stabilization of Linear
Systems
212 pp. 1993 [3-540-19862-8]

Vol. 192: Smith, R.S.; Dahleh, M. (Eds)
The Modeling of Uncertainty in Control
Systems
412 pp. 1993 [3-540-19870-9]

Vol. 193: Zinober, A.S.I. (Ed.)
Variable Structure and Lyapunov Control
428 pp. 1993 [3-540-19869-5]

Vol. 194: Cao, Xi-Ren
Realization Probabilities: The Dynamics of
Queuing Systems
336 pp. 1993 [3-540-19872-5]

Vol. 195: Liu, D.; Michel, A.N.
Dynamical Systems with Saturation
Nonlinearities: Analysis and Design
212 pp. 1994 [3-540-19888-1]

Vol. 196: Battilotti, S.
Noninteracting Control with Stability for
Nonlinear Systems
196 pp. 1994 [3-540-19891-1]

Vol. 197: Henry, J.; Yvon, J.P. (Eds)
System Modelling and Optimization
975 pp approx. 1994 [3-540-19893-8]

Vol. 198: Winter, H.; Nüßer, H.-G. (Eds)
Advanced Technologies for Air Traffic Flow
Management
225 pp approx. 1994 [3-540-19895-4]

Vol. 199: Cohen, G.; Quadrat, J.-P. (Eds)
11th International Conference on
Analysis and Optimization of Systems –
Discrete Event Systems: Sophia-Antipolis,
June 15–16–17, 1994
648 pp. 1994 [3-540-19896-2]

Vol. 200: Yoshikawa, T.; Miyazaki, F. (Eds)
Experimental Robotics III: The 3rd
International Symposium, Kyoto, Japan,
October 28-30, 1993
624 pp. 1994 [3-540-19905-5]

Vol. 201: Kogan, J.
Robust Stability and Convexity
192 pp. 1994 [3-540-19919-5]

Vol. 202: Francis, B.A.; Tannenbaum, A.R. (Eds)
Feedback Control, Nonlinear Systems,
and Complexity
288 pp. 1995 [3-540-19943-8]

Vol. 203: Popkov, Y.S.
Macrosystems Theory and its Applications:
Equilibrium Models
344 pp. 1995 [3-540-19955-1]

Vol. 204: Takahashi, S.; Takahara, Y.
Logical Approach to Systems Theory
192 pp. 1995 [3-540-19956-X]

Vol. 205: Kotta, U.
Inversion Method in the Discrete-time
Nonlinear Control Systems Synthesis
Problems
168 pp. 1995 [3-540-19966-7]

Vol. 206: Aganovic, Z.;.Gajic, Z.
Linear Optimal Control of Bilinear Systems
with Applications to Singular Perturbations
and Weak Coupling
133 pp. 1995 [3-540-19976-4]

Vol. 207: Gabasov, R.; Kirillova, F.M.;
Prischepova, S.V.
Optimal Feedback Control
224 pp. 1995 [3-540-19991-8]

Vol. 208: Khalil, H.K.; Chow, J.H.;
Ioannou, P.A. (Eds)
Proceedings of Workshop on Advances
inControl and its Applications
300 pp. 1995 [3-540-19993-4]

Vol. 209: Foias, C.; Özbay, H.;
Tannenbaum, A.
Robust Control of Infinite Dimensional Systems:
Frequency Domain Methods
230 pp. 1995 [3-540-19994-2]

Vol. 210: De Wilde, P.
Neural Network Models: An Analysis
164 pp. 1996 [3-540-19995-0]

Vol. 211: Gawronski, W.
Balanced Control of Flexible Structures
280 pp. 1996 [3-540-76017-2]

Vol. 212: Sanchez, A.
Formal Specification and Synthesis of
Procedural Controllers for Process Systems
248 pp. 1996 [3-540-76021-0]

Vol. 213: Patra, A.; Rao, G.P.
General Hybrid Orthogonal Functions and
their Applications in Systems and Control
144 pp. 1996 [3-540-76039-3]

Vol. 214: Yin, G.; Zhang, Q. (Eds)
Recent Advances in Control and Optimization of
Manufacturing Systems
240 pp. 1996 [3-540-76055-5]

Vol. 215: Bonivento, C.; Marro, G.;
Zanasi, R. (Eds)
Colloquium on Automatic Control
240 pp. 1996 [3-540-76060-1]

Vol. 216: Kulhavý, R.
Recursive Nonlinear Estimation: A Geometric
Approach
244 pp. 1996 [3-540-76063-6]

Vol. 217: Garofalo, F.; Glielmo, L. (Eds)
Robust Control via Variable Structure and
Lyapunov Techniques
336 pp. 1996 [3-540-76067-9]

Vol. 218: van der Schaft, A.
L_2 Gain and Passivity Techniques in Nonlinear
Control
176 pp. 1996 [3-540-76074-1]

Vol. 219: Berger, M.-O.; Deriche, R.;
Herlin, I.; Jaffré, J.; Morel, J.-M. (Eds)
ICAOS '96: 12th International Conference on
Analysis and Optimization of Systems - Images,
Wavelets and PDEs:
Paris, June 26-28 1996
378 pp. 1996 [3-540-76076-8]

Vol. 220: Brogliato, B.
Nonsmooth Impact Mechanics: Models,
Dynamics and Control
420 pp. 1996 [3-540-76079-2]

Vol. 221: Kelkar, A.; Joshi, S.
Control of Nonlinear Multibody Flexible Space
Structures
160 pp. 1996 [3-540-76093-8]

Vol. 222: Morse, A.S.
Control Using Logic-Based Switching
288 pp. 1997 [3-540-76097-0]